ALSO FROM *NEW SCIENTIST*

Why Don't Penguins' Feet Freeze?
And 114 Other Questions

Does Anything Eat Wasps?
And 101 Other Unsettling, Witty Answers
to Questions You Never Thought You Wanted to Ask

Stroke a Martian:
And 99 Other Things to Do Before You Die

100 Things to Do Before You Die
(Plus a Few to Do Afterwards)

How to Fossilize Your Hamster

How to Fossilize Your Hamster

And Other Amazing Experiments for the Armchair Scientist

MICK O'HARE

A HOLT PAPERBACK

Henry Holt and Company • New York

Holt Paperbacks
Henry Holt and Company, LLC
Publishers since 1866
175 Fifth Avenue
New York, New York 10010
www.henryholt.com

A Holt Paperback® and ® are registered trademarks of
Henry Holt and Company, LLC.

Anyone undertaking any of the experiments described in this book does so
entirely at their own risk. Children must always be supervised by a responsible
adult. General warnings are given throughout the text, but whether or not a
warning is given, it is the responsibility of the reader always to take appropriate
precautions and to exercise due care. No responsibility for any consequences,
however caused, will be accepted by the author, *New Scientist*, Profile Books, Henry
Holt and Company, or any of the individuals and organizations acknowledged for
their assistance or contributions.

Originally published in Great Britain in 2007
by Profile Books Ltd., London

Library of Congress Cataloging-in-Publication Data
O'Hare, Mick, 1964–
  How to fossilize your hamster : and other amazing experiments for the
armchair scientist / Mick O'Hare.
     p.   cm.
  Includes index.
  ISBN-13: 978-0-8050-8770-3
  ISBN-10: 0-8050-8770-2
 1. Science—Experiments. 2. Science—Miscellanea. I. Title.
  Q164.O43 2008
  507.8—dc22   2007031876

Henry Holt books are available for special promotions and
premiums. For details contact: Director, Special Markets.

First Holt Paperbacks Edition 2008

Designed by Kelly S. Too
Text illustrations by Frazer Hudson
Margin illustrations by Brett Ryder

Printed in the United States of America
3   5   7   9   10   8   6   4   2

CONTENTS

◯ INTRODUCTION

Experiments are what make science tick. Observing, recording, and observing again have taught us everything we know about our universe and the world around us. Without experimental evidence, science is reduced to little more than a set of theories. Real scientists—from Newton and the apple to Pavlov and his dogs—get out there and observe. And when scientists have observed and recorded what they have seen, they repeat the experiment to ensure its results are verifiable.

Which is what this book is all about—experimenting and seeing for yourself. It's where the real pleasure in science lies. Science isn't restricted to the lab; it's experimenting wherever you are and using whatever is on hand. By reading this book, you'll understand how great science has been achieved through experimentation.

We focus on how the trivial reveals the eye-opening, a *New Scientist* trademark. You can find out why shaken and stirred martinis taste different, but not why the universe is expanding. You'll learn how to extract iron from your breakfast cereal, but you won't discover what's inside a black hole. The big stuff can come later, after you've been inspired by the experiments here. Best of all, you'll be able to ensure that your dear departed hamster is preserved for eternity (though you won't be able to discover just how long eternity is).

While not every entry in the book can be termed an experiment, each allows you to try out simple things at home that demonstrate science in action. These range from the chemical (How do cola and Mentos make such an explosive mix?) to the biological (Why do some things pass straight through the digestive system apparently untouched?). And at *New Scientist*, we've unearthed a preponderance of experiments featuring alcoholic drinks—the magazine is staffed by journalists, after all. While these are for adults only, the rest are geared to be enjoyed by the entire family. Children should be supervised when carrying out any of the experiments we describe. For although they have been chosen with safety in mind, a few involve potentially hazardous materials (for example, hot water, matches, and knives), so care needs to be taken.

Some of the experiments will almost certainly need a large outdoor space to try out, but you should be able to attempt most of them as soon as you get this book home or shortly after a trip to the local store. Hardly any require specialized equipment. If you need further instruction on how to make them work, you'll find some of them on our Web site: http://www.new scientist.com/hamster.

The best science is always collaborative, so I am grateful to all the readers of *New Scientist*, colleagues, scientists, and smart people everywhere who have helped make this book possible—a roll call appears in the acknowledgments.

Finally, remember: scientists' theories are there to be challenged. If you carry out the experiments in this book and draw different conclusions, let us know—you will find our contact details in the acknowledgments, too. Science is constantly evolving in response to new evidence; that is what makes it so wonderful and enlightening. There's nothing we'd enjoy more than being proved wrong by someone who has carried out an experiment from this book and come to a different conclusion. After all, real scientists always experiment, not once, not twice, but again and again. . . .

MICK O'HARE

How to Fossilize Your Hamster

1 IN THE KITCHEN

○ PLASTIC MILK

Can you make plastic in your kitchen?

You would imagine that you'd need some pretty noxious, smelly chemicals to make plastic, but you can actually find the things you need to make malleable, doughy pieces of it in your own home. Instead of putting vinegar on some fish and chips and crying over your spilled milk, use them to become a polymer chemist.

WHAT DO I NEED?
◆ a pint of milk
◆ a saucepan
◆ a sieve or colander
◆ a spoon for stirring
◆ four teaspoons of white vinegar
◆ rubber gloves
◆ water

WHAT DO I DO? Pour the milk into the pan and gently warm it. When the milk is simmering (don't let it boil), stir in the white vinegar until you notice whitish yellow rubbery lumps beginning to curdle in the mixture and the liquid clearing. Turn off the heat and let the pan cool.

WHAT WILL I SEE? First of all you'll smell the vinegary reaction, which is the key to the process at work. As the vinegar is added and stirred, the liquid gets clearer, and the yellowy rubbery lumps form. When the pan has cooled you can sieve the lumps from the liquid, pouring the liquid down the sink. Put on the rubber gloves and wash the lumps in water. You can then press them together into one big blob—they will be squishy and will feel as if they are going to fall apart, but they will stick together after some firm kneading. You can now use your artistic skills to fashion the material into the shapes of your choice—the *New Scientist* staff came up with balls, stars, a heart shape for a pendant, and even dinosaur footprints. Leave the material to dry for a day or two, and it will become hard and plastic enough to paint and varnish.

WHAT'S GOING ON? You have used the combination of an acid—in this case vinegar, which contains acetic acid—and heat to precipitate casein (a protein) from the milk. Casein is not soluble in an acid environment, and so when the vinegar is added to the warm milk, it appears in the form of globular plastic-like lumps. Casein behaves like the plastics we see in many objects around us, such as computer keyboards and phones, because it has a similar molecular form. The plastics in everyday objects are based on long-chain molecules called polymers. These are of high molecular weight and get their strength from the way their billions of interwoven crisscrossing molecules tangle together.

P.S. Some forms of cheese making rely on a similar technique—the name "casein" comes from *caseus*, the Latin for cheese. The Indian cheese known as *paneer* is made in a very similar way to the plastic you have just made, although in this case lemon juice is the acid used rather than vinegar. Afterward, unlike our plastic milk, it is not dried out and allowed to harden to tooth-breaking consistency, and so remains soft and edible.

☯ IRON MAN

Breakfast cereals often claim to be fortified with iron. Well, are they?

They are and, more amazingly, if you have a magnet you can extract it, too! So ponder the ingredients list on your box of cornflakes while you are munching breakfast, and then set about removing one of them.

WHAT DO I NEED?
- breakfast cereal fortified with iron (cornflakes work, but check on the side of the box to see what the iron content is—the higher, the better)
- a plastic cup
- a spoon or pestle to crush the cereal (better still, a blender)
- hot water
- a very strong magnet

- clean white paper
- a clear, sealable plastic sandwich bag

WHAT DO I DO? Fill the cup to about two-thirds full with cereal, and with the spoon or pestle crush the cereal into a fine powder. It is worth spending a lot of time on this stage—the finer the powder, the better. Put the crushed cereal into the sandwich bag and add hot water. Leave the mixture for about fifteen to twenty minutes. Now gently tilt the bag forward so that the cereal collects on one side, and place the magnet along the outside of the bag near the cereal, running it over the bottom, because the iron tends to sink. Tilt the bag back so that the cereal runs away from the magnet. You can also lay the bag flat on the table and stroke the magnet across it toward a single corner of the bag.

Alternatively, if you are using a blender, put the cereal straight into the blender and add hot water until the cereal is submerged. Wait for about fifteen to twenty minutes until the cereal is soft, then blend it all together. While the blender is whirring, place the magnet on the outside of the blender near the mixture, and keep it there as you turn the blender off.

WHAT WILL I SEE? The magnet will attract a black fuzz of iron. Move the magnet over the surface of the bag or blender and the tiny pieces of iron will follow it.

WHAT'S GOING ON? The black stuff really is iron in your cereal—the same stuff that is found in nails and trains and motorbikes. And it's quite heavy, which is why you need to make sure you run your magnet along the bottom of the bag. The iron is added to the mix when the cereals are being made, and you really do eat it when you devour your cornflakes.

The reason it is added in a form that you can extract is that iron ions (iron that would more easily combine with other molecules in the cereal) increase the spoilage rate of the food. Using iron in its pure metal form gives the cereal a longer shelf life.

The hydrochloric acid and other chemicals in your stomach dissolve some of this iron, and it is absorbed through your digestive tract, although much of it remains untouched and is excreted.

P.S. Humans need iron for many important bodily functions. Red blood cells carry hemoglobin, of which iron is a key constituent. Hemoglobin transports oxygen through the blood to the rest of the body by binding oxygen to its iron atom and carrying it through the bloodstream. As our red blood cells are being replaced constantly, iron is an essential part of our diet.

◎ FIZZ FALLACY

Can you maintain the fizz in champagne overnight if a teaspoon is suspended in the neck of the bottle?

I have included this experiment not because it is spectacular or even because it actually works, but to illustrate the use and importance of controlled conditions when attempting to prove or disprove a theory. For any experiment in this book, you can add a control for comparison to test whether a particular ingredient is the cause of the reaction you are observing.

WHAT DO I NEED?

- at least two bottles of champagne
- a fridge
- a teaspoon
- champagne flutes

WHAT DO I DO? Open two bottles of champagne. Drink some from each, then place a teaspoon into the neck of one of the bottles, with the handle dangling downward and with no part of the teaspoon touching the liquid. Drink a little more if it is. The other bottle should be left open. To maintain a true control, keep the amount of champagne in each bottle equal. Now place both in the fridge and leave them overnight. Test them at regular intervals, noting how fizzy each bottle is and whether there is any significant difference between the two. I suggest you test them the following morning, lunchtime, and evening, and again on subsequent days, until the champagne in both bottles has lost all its fizz.

WHAT WILL I SEE? You'll need to be objective in your assessment of the fizz quotient of each bottle, but you will find, especially if you repeat the experiment as any decent scientist would, that both bottles are equally fizzy at each stage of testing. As an objective measure you could see how much champagne needs to be added to the champagne flutes for the bubbles to reach the rim, but for this to work you'll have to make sure you pour all glasses at the same rate.

WHAT'S GOING ON? This question is a classic example of being misled by uncontrolled experiments.

Because people rarely have two bottles open at the same time, if they have stored an unfinished bottle with a suspended spoon in its neck they attribute the following day's unexpected longevity of fizz to the spoon. However, in your tasting experiments you will have found that the truth of the matter is that champagne—surprisingly—keeps its fizz for three days or more, with or without the teaspoon.

The opened bottle without the teaspoon acts as a control against which you are able to gauge the fizziness of the bottle with the teaspoon in its neck. Both bottles decrease in fizziness at exactly the same rate.

It's not uncommon to attach significance to apparently linked events when there is no control data to compare them with. Often you'll hear people saying things like, "How incredible, I was just thinking of you when the phone rang and it was you." Telepathy is not involved here. We merely ignore the huge number of times we think of people and the phone does not ring.

P.S. If your budget doesn't stretch to champagne, the experiment will work just as well with cava, Asti Spumante, or a domestic sparkling wine.

⊃ PASTA PUZZLE

If a strand of dried spaghetti is held at both ends and bent, why does it always break into three or more pieces?

This is, indeed, a strange phenomenon. Surely holding a strand of dried spaghetti at both ends and bending it until it breaks should produce just two pieces, but it hardly ever does—usually three or even more pieces are the result. This conundrum first appeared in *New Scientist* in 1995 and was repeated in 1998. Even so, we didn't get to the bottom of it until 2006. It's a problem that has taxed greater minds than ours, including that of the Nobel Prize–winning physicist Richard Feynman.

WHAT DO I NEED?
- strands of dried spaghetti
- something to catch them in

WHAT DO I DO? Hold a strand of spaghetti at both ends and bend it until it breaks. Count the pieces. Repeat with the other strands.

WHAT WILL I SEE? In nearly all cases the spaghetti will break into three or more pieces. Even on the rare occasions when it seems to break into only two you'll often find a stray shard or splinter has flown off into the nether regions of your kitchen.

WHAT'S GOING ON? Back in 1998, a *New Scientist* reader tried to come to grips with the issues involved and came very close to solving the problem.

First, when you bend a piece of spaghetti, it does not usually break at the apex of the bend where the stresses are highest, because failure in the spaghetti is controlled by defects in the pasta. The first break occurs at a point

1. spaghetti

2. both strands subject
to flexual waves
after snapping

3. flexual wave causes
spaghetti to spring
back after breaking;
already weakened here
by previous bending

4. third piece snaps off
after springing back

near the apex where the combination of stress level and defect size reaches a critical value. This breaks the original piece into a long and a short piece. After the break, as the longer piece snaps back, the whipping action sends the tip beyond the neutral point (the original straight state of the piece of spaghetti) and activates the next defect on what was the long side. This defect has already been opened up on the outside of the curved spaghetti by the first bending, so it doesn't take much to finish off the crack by bending it in the other direction.

Second, the sequence of events can be determined by looking at the broken ends of the spaghetti pieces. When a break occurs, the fracture starts cleanly on the stretched convex side and ends slightly raggedly on the compressed concave side where a small splinter—or spicule—is usually torn away from one side of the break. Additionally, careful inspection of the ejected middle piece of spaghetti will reveal evidence of spicule formation at both ends and that these are on opposite sides. This shows that the two breaks that generate the middle piece occur while the spaghetti is bending in opposite directions, which is consistent with the dynamics of linear spaghetti structures.

While being very much on the right track, the reader's observations only partly answered the conundrum. It took Basile Audoly and Sébastien Neukirch to verify what was going on in their paper "Fragmentation of Rods by Cascading Cracks: Why Spaghetti Does Not Break in Half," published in *Physical Review Letters* (volume 95, number 9, December 22, 2004, page 95505).

Audoly and Neukirch broke strands of spaghetti of varying thicknesses and lengths by clamping one end and bending them from the other. They found that the unexpected three-part breakage occurs because of what are known as flexural waves. When the curvature of the spaghetti reaches a critical point, the first break appears. The shock of this causes a flexural wave to ripple down each of the two resulting lengths of pasta at high speed and amplitude.

The two halves formed by the initial break do not have time to relax and straighten before being hit by the flexural wave, which causes them to curve even further and suffer more breaks, leading to a cascade of cracks in the pasta. Often more than three pieces are created when this happens.

While spaghetti snapping is in itself a rather humdrum if fun pastime, Audoly and Neukirch's work also provides important information about failures in other elongated, brittle structures, including human bones and bridge spans.

P.S. In *No Ordinary Genius*, the illustrated biography of Richard Feynman published in 1994, Danny Hills describes his and Feynman's experiments with spaghetti: "If you get a spaghetti stick and break it, it turns out that instead of breaking in half, it will almost always break into three pieces. Why is this true—why does it break into three pieces? Well, we ended up at the end of a couple of hours with broken spaghetti all over the kitchen and no real good theory about why spaghetti breaks into three." This seems to have been a common occurrence—apparently visitors to Feynman's home were often presented with sticks of spaghetti and asked to help solve the problem.

There is, therefore, a delicious irony in the fact that while this puzzle drove Feynman, a Nobel Prize winner, in physics, to distraction, those who discovered the reason why it happened were awarded the antithesis of Nobel fame, the Ig Nobel Prize for Physics in 2006, forty-one years after Feynman won his Nobel Prize. Nobel Prizes are awarded for supreme achievements in scientists' chosen fields, while Ig Nobels, from the opposite end of the research spectrum, are awarded for success in the areas of improbable research, humor, and, quite often, silliness.

WANT TO READ MORE? A video of breaking spaghetti can be seen at http://www.lmm.jussieu.fr/spaghetti/index.html, where you can also find out more about Audoly and Neukirch's research.

No Ordinary Genius (1994) was edited by Christopher Sykes and published by W. W. Norton & Company.

⊙ BANANA ARMOR

What's the best way to store bananas to keep their skins from turning brown quickly?

This is counterintuitive to those of us brought up to believe that chilling foodstuffs slows decay, but a simple experiment will show us if it's true or not.

WHAT DO I NEED?

◆ two or more bananas (and possibly some fresh banana skins)

- a fridge (or a domestic freezer)
- lemon juice

WHAT DO I DO? Place one banana in the fridge and leave the other at room temperature (approximately 68°F). Observe each banana three or four times a day and note the relative discoloration of the skins. As a side experiment, rub a third banana with lemon juice before subjecting it to the fridge conditions.

WHAT WILL I SEE? The banana in the fridge will brown or blacken faster than the one at room temperature. However, a banana rubbed with lemon juice and placed in the fridge will not decay at the same rate as the untreated one.

WHAT'S GOING ON? While many fruits are stabilized by refrigeration, most tropical and subtropical fruits, and bananas in particular, exhibit chill injury. Tests show that the ideal temperature for long-term banana storage is 56°F. Below 50°F, spoilage is accelerated because their cells' internal membranes are damaged, releasing enzymes and other substances. Banana skin can blacken overnight as it softens and breaks down.

The membranes that keep separate the contents of the various compartments inside a cell are essentially two layers of slippery fat molecules or lipids. Chill these membranes and the molecules get more sticky, making the membranes less flexible. Bananas adjust the composition of their membranes to give the right degree of membrane fluidity for the temperature at which they normally grow. They do this by varying the amount of unsaturated fatty acids in the membrane lipids: the greater the level of unsaturated fatty acids, the more fluid the membrane at a given temperature. If you chill the fruit too much, areas of the membrane become too viscous, and it loses its ability to keep the different cellular compartments separate. Enzymes and substrates that are normally kept apart therefore mix as the membranes collapse and hasten the softening of the fruit flesh.

Skin blackening involves the action of a different enzyme from those involved in flesh softening. In the skin, polyphenol oxidase breaks down naturally occurring phenols in the banana skin into substances similar in structure to the melanin found in suntanned human skin. So the browning starts sooner in refrigerated bananas because of chill-induced membrane damage that allows the normal process of decay—which would have occurred anyway at room temperature—to begin earlier. The cold itself does not speed up the browning part of the reaction. Indeed, if chill damage in a fridge is induced first, removing the banana from the fridge then hastens the process as the reaction that causes the browning, once it is under way, is speeded up by heat.

This can be demonstrated by putting a banana skin in a freezer for a few hours. The inner surface will stay creamy white because, although the membranes are destroyed by the freezing process, the oxidases cannot work at such low temperatures. Then let it thaw overnight at room temperature. In the morning it will be pitch-black due to the damage the cell membranes suffered in the freezer. Yet had the cold itself caused the blackening, it would have turned dark while it was being frozen.

Decay can be slowed by acids, which prevent the release of the polyphenol oxidase enzyme. This is why adding lemon juice—which is rich in citric acid—to skins can slow the browning process. A similar slowing of the breakdown process can be seen if bananas are coated in wax; this prevents oxygen from reaching the skin and speeding up decay.

P.S. In commercial storage of tropical fruit, chilling injury is a big problem. Unlike temperate fruits such as apples and pears, which can be stored at temperatures close to freezing, tropical fruits break down faster in the cold. Because tomatoes, now commonplace in northern Europe and North America, are actually a semitropical fruit, the evidence suggests that they will last longer out of the fridge than in it. I look forward to hearing from readers about their home experiments on tomatoes and any number of other fruits whose growing environments straddle temperate and tropical climes.

☉ BOUNCING RICE

Why does a grain of cooked rice in a glass of lemon-lime soda repeatedly rise to the surface and fall to the bottom?

I have fond memories of fooling gullible schoolmates with this "magic" trick, telling them that for a modest sum

I could make objects at the bottom of the glass mystically rise to the surface. That was until I was outed by the head of chemistry and forced to hand back my ill-gotten proceeds.

WHAT DO I NEED?
- a glass of lemon-lime soda (other fizzy drinks will work just as well, but transparent drinks are better than opaque ones if you want to see the rice)
- grains of cooked and uncooked rice

WHAT DO I DO? Drop a grain of cooked rice into the soda and wait. For comparison, put a grain of uncooked rice in a glass of soda. As a control (discussed in the "Fizz Fallacy" experiment on page 5), put a grain of cooked rice in a glass of still water.

WHAT WILL I SEE? The cooked rice will sink to the bottom of the glass of soda. Then, after a short while, it will begin an inexorable cycle of rise and fall between the surface and the glass bottom. The uncooked rice will also rise and fall, but more slowly. The rice in the control will do nothing.

WHAT'S GOING ON? Initially, the grain of rice sinks in the soda because it is denser than the liquid around it. However, once the grain has reached the bottom of the glass, bubbles of carbon dioxide begin to collect around it. This is because bubbles in fizzy drinks tend to form preferentially around rough surfaces, which provide what are known as nucleation sites. Nucleation sites are hydrophobic (they repel water) and allow gas pockets

Bubbles form on rice grain in lemon-lime soda.

Buoyant rice grain rises to surface.

Bubbles pop at surface and rice grain sinks again to start the process once more.

to form without first forming the tiny bubbles. The uncooked grain has fewer rough edges and fewer nucleation sites, so the process takes longer.

Eventually, the bubbles of carbon dioxide begin acting as if they were airbags inflated underwater and attached to the grain. The rice then becomes buoyant and starts rising. When it reaches the surface, the bubbles burst and it sinks again, restarting the process. If you leave the soda and rice for long enough, the drink will go flat and the rice will stop ascending.

P.S. The *New Scientist* team is indebted to the students of the Aberdeen University Scottish Country Dance Society who were kind enough to send in their extensive research in this field carried out over a number of years in controlled laboratory conditions (the student bar) using a variety of drinks and bar snacks. They tell us that the most consistent effects were usually obtained using salted peanuts and cheap lager. In fact, a handful of peanuts rising and falling in a pint of lager can produce an effect similar to the lava lamps popular in the 1960s and 1970s. Unfortunately, but not surprisingly, the salt on the nuts makes cheap lager unpalatable, so do this only if you have no intention of consuming the beer later.

Interestingly, Newcastle Brown Ale and many bottled ciders are so fizzy that bubbles form too quickly, and the salted peanuts never have time to sink back to the bottom but remain at or near the surface. Dry-roasted peanuts are far too dusty to work properly, and potato chips and pork rinds don't sink in the first place but float soggily at the surface.

⟲ GREEN EGGS AND CABBAGE

Why does the juice from cooked red cabbage turn fried egg whites green?

Chef Heston Blumenthal, one of the founding fathers of molecular gastronomy, would love this one. So would Dr. Seuss. You need only a small amount of red cabbage juice to change the color of fried eggs from white to green. Your kids and brunch guests will be completely taken aback.

WHAT DO I NEED?
* shredded red cabbage
* water
* a pot
* a jug
* oil
* a frying pan
* one egg

WHAT DO I DO? Boil the cabbage in a pot for twenty minutes and let it cool. Squeeze the juice from the cooled cabbage into a jug. Heat the oil in a frying pan and begin to fry the egg until the white is just turning from clear to white. Drip a small amount of cabbage juice into the setting egg white.

WHAT WILL I SEE? The egg white will turn lurid green where the juice hits it.

WHAT'S GOING ON? Red cabbage juice is a good indicator of whether a substance is an alkali or an acid.

If added to an alkali, such as ammonia, it will turn green; if added to an acid, such as lemon juice, it will turn red. In neutral substances it is purple, the natural color of red cabbage. Because egg white (mostly the protein albumen) is alkaline, it turns green. Any number of substances can be tested in this way, although take care to avoid strongly corrosive chemicals such as drain cleaners or bleach, because these can be dangerous.

The experiment works because red cabbage contains water-soluble pigments called anthocyanins (also found in plums, apple skins, and grapes). These change color depending on whether they are in the presence of an acid or an alkali. These change the number of hydrogen ions attached to the molecule—acids donate hydrogen ions while alkalis remove them—and it is the presence or absence of hydrogen ions that is responsible for creating the different colors. This explains why red cabbage that is pickled turns red rather than its natural purple. Pickling takes place in vinegar, which is acidic.

Red cabbage juice breaks down quite quickly, so if you are going to use it to test the acidity or alkalinity of other household foods or products, use it sparingly and fast.

P.S. A trick that will prove popular with children is to create "magic" paper. Soak cheap absorbent paper in boiled red cabbage water and leave it to dry. Then paint it with household substances such as vinegar, orange juice, or powdered clothes detergent dissolved in water. A range of colors will appear on the paper depending on the acidic or alkaline nature of the paints used.

☺ BURNT OFFERINGS

Why does brown or whole wheat bread char more quickly than white bread when you toast it? (Which may explain why I burn the toast but my wife doesn't. . . .)

WHAT DO I NEED?
- slices of brown or whole wheat bread
- slices of white bread
- a toaster

You may also want to have butter and jam or jelly handy to ensure your efforts are not wasted.

WHAT DO I DO? Place the brown bread in the toaster. Check it every fifteen seconds to see when it starts to burn (you'll need to set a clear if subjective measure of what you consider to be charred bread in advance). Take a note of the time. Repeat with the white bread.

WHAT WILL I SEE? You'll notice that the brown or whole wheat bread toasts much faster than the white bread.

WHAT'S GOING ON? During heating, a complex reaction occurs between the proteins and sugar contained in the bread. This is known as the Maillard reaction, and it produces the typical flavor we know from toasted bread as well as the color formed as the bread toasts.

The Maillard reaction is a chemical one between an amino acid and a reducing sugar. It usually requires the addition of heat, as in the case of toasting bread. The reaction is widely used in the food industry, producing different flavors and odors according to the type of amino acid involved in the reaction.

Because brown and whole wheat bread contain more sugar and protein than white bread, they undergo more rapid Maillard browning than white bread does.

One other factor may help to explain why white bread chars more slowly: albedo, the proportion of incident light or radiation that is reflected by a surface. White bread reflects more radiation than brown bread, which is why it appears whiter. Because darker breads absorb more radiation in the form of heat from the toaster, they heat up and burn more quickly.

P.S. While you have the loaves of bread out, you can try another experiment. Take a fresh, untoasted slice of bread and spread it with honey. Then set it aside. If you leave it for a few minutes, you'll see that it becomes concave on the side spread with honey. This is because bread is approximately 40 percent water while honey is a strong solution containing approximately 80 percent sugar. Sugar is hygroscopic, which means that it soaks up moisture. This causes moisture to be drawn out of the bread and into the honey by osmosis. Extracting the water makes the bread shrink, but only on the side exposed to the honey. This explains why the bread becomes concave.

This is less likely to happen if you butter your bread before spreading on the honey. Butter forms a fat-rich, water-impermeable layer that protects the bread from dehydration by the honey.

WANT TO READ MORE? The chemistry of different bread types is discussed in *The Composition of Foods* by Robert

Alexander McCance and Elsie Widdowson (Elsevier/North-Holland Biomedical Press).

○ STRINGS ATTACHED

Most hard cheeses such as cheddar go stringy when grilled, yet Indian paneer *and Greek-Cypriot* halloumi *don't melt at all; they just char while maintaining their shapes. Why is this?*

All cheese is made from milk, so that obviously cannot be the reason. When you try this out, keep a slice of toast from the "Burnt Offerings" experiment (page 21) on hand—there's no point in wasting melted cheese.

WHAT DO I NEED?
- cubes of cheddar
- cubes of *paneer* or *halloumi* cheese
- kebab skewers
- an oven broiler
- bread to eat the results with

WHAT DO I DO? Thread the cheese cubes onto the skewers, place under the broiler, heat at a high temperature, and watch what happens.

WHAT WILL I SEE? The cheddar will start to go stringy and drip into the broiler's pan. (This is where some bread, slowly toasting below the cheese skewer, will come in handy.) Meanwhile, the *paneer* or *halloumi* will retain its cube shape and often char. Fortunately, it, too, can be eaten after it has cooled a little.

WHAT'S GOING ON? Uncooked cheddar contains long-chain protein molecules curled up in a fatty, watery mess. It also has a low melting point which means it becomes runny before it burns. As the fats and proteins melt and drop under the weight of gravity, they are dragged into dripping strings—just as they are if you take a bite from a grilled cheese sandwich or from the bread that has collected the drippings from your cheddar skewer. As the cheese softens, the long-chain protein molecules unravel and form fibers. From eating pizza, you'll know that mozzarella displays this quality even better and can string out to a foot or more. In fact, string length is a measure of protein content—the longer the string of cheese, the more protein it contains.

However, in addition to melting and becoming stringy, the protein in cheddar is denatured by the heat. This is why previously molten cheddar turns into an unpleasant rubbery lump after it cools. *Paneer, halloumi,* and similar cheeses are produced by taking advantage of this process. They have already been partially cooked—some with heat and others with exposure to acid—while they are made, so important changes have already taken place before you put them under your grill. The denaturing effect means the cheese is already in a rubbery state and so keeps its shape. That ensures that the *paneer* in Indian dishes doesn't ooze all over the place and makes *halloumi* perfect for use in kebabs. Of course, these cheeses can still burn easily, so keep a close eye on this experiment at all times.

P.S. Try reheating cold, previously molten cheddar to see whether its properties have become like those of *paneer* and *halloumi*. You'll find you've produced your own denatured cheese in the kitchen, even though it's just about inedible.

9 IN THE DUMPS

Why do matzo balls and gnocchi dumplings float while uncooked ones remain close to the bottom of the pot?

I have happy memories of my mom's home-cooked beer and beef stew with dumplings, which has flour-based dumplings not unlike matzo balls and gnocchi, so this question offered me the perfect opportunity to revisit a favorite childhood dish. If you don't have a family recipe for matzo balls or gnocchi that you'd like to cook up to test this principle, I've provided a recipe for dumplings below, courtesy of Mom, who has always been an expert at traditional English fare.

WHAT DO I NEED?
- uncooked matzo balls, gnocchi, or other dough-based dumplings
- a pot of simmering liquid of your choice (a stew or soup stock works well for converting your research into a sublime culinary experience)
- a slotted spoon

WHAT DO I DO?
Lower the uncooked dumplings carefully into the simmering liquid with a slotted spoon.

WHAT WILL I SEE?
The uncooked dumplings will initially sink to the bottom of the pan. As they cook, they

will rise to the surface. While I admit that this is easier to see if you are using boiling water, it's more efficient to cook them in a stew or soup since, of course, you'll end up with dinner after the experiment has been completed.

WHAT'S GOING ON? Two factors are at work. It seems the main reason the dumplings float once they are cooked is because bubbles become trapped in their dough matrix—held together with the gluten in the flour or meal—and provide buoyancy. Similar foodstuffs, such as meatballs made from minced meat, egg, bread crumbs, and herbs, have the same propensity. First they sink, then they rise. This is because the air spaces bound up in the dough matrix expand when they are placed in boiling liquid, making the mixture less dense. This is what makes them buoyant. Consequently, if left to cool, various dumplings and meatballs will return to the bottom of the pan as the air spaces shrink, so the golden rule is eat them while they are hot. And they taste far better that way.

There is, however, another factor at work, especially in the case of flour-based dumplings like the ones in my mom's stew, and that is the presence of baking powder. Most varieties of this style of dumplings are made with plain flour and a large quantity of baking powder or with self-rising flour. Baking powder is a leavening agent containing an alkali, most commonly sodium bicarbonate, and an acid plus starch to prevent the mixture from becoming damp. When it is dissolved in water, the alkali reacts with the acid to produce carbon dioxide, which creates bubbles in the dumplings. Self-rising flour has a similar effect, although in this case the leavening agent is already part of the mixture (so there is no need to add baking powder).

Try making two batches of dumplings, one with a leavening agent and one without, and you'll find the ones without leavening are far more reluctant to float. Given enough time and heat, though, the air spaces in their dough matrix will ensure they eventually reach the surface.

P.S. This recipe for dumplings has been in the O'Hare family for generations. We cannot claim its superiority over other dumpling recipes, but I do know that it leads to dumplings that follow the experimental rules outlined above. First they sink, then they float.

> 1 cup self-rising flour
> 5 tablespoons plus 1½ teaspoons shredded suet (you can use a vegetarian version if you prefer, or substitute solid vegetable shortening)
> Decent-sized pinches of salt and pepper
> 1 teaspoon dried herbs
> Approximately 3 tablespoons of water

Sift the flour into a bowl and mix it well with the suet, salt, pepper, and herbs. Add the water a little at a time to make a slightly soft but not gooey dough. Lightly flour your hands and form your dough mixture into six to eight balls. Using a slotted spoon, carefully place them in a simmering stew or soup of your choosing (or water) over medium to medium-high heat and allow them to cook for twenty-five to thirty minutes. Check them after twenty minutes—if the cooking liquid is boiling vigorously, they'll be cooked sooner.

If you have family recipes of your own, you can compare and contrast the buoyancy or lack thereof of your own dumplings with the O'Hare version.

○ SHAPE SHIFTING

Why is cornstarch mixed with cooking oil attracted to a balloon?

With its very high proportion of starch, cornstarch gives rise to a number of weird effects that make it the perfect ingredient for use in scientific demonstrations. In addition to showing off the properties of starch molecules, this experiment shows the strength of even very small electrical charges. (For another experiment involving charges, see "Wayward Water" in chapter 5.)

WHAT DO I NEED?
◆ cornstarch
◆ cooking oil
◆ an inflated balloon
◆ a wooden spoon (the spoon must be wooden because you need an electrical insulator)

WHAT DO I DO? Mix the cornstarch and oil together until the mixture resembles a thick cream. Rub the balloon on your clothing to create a charge on its surface (as in the old party trick of rubbing a balloon and sticking it to the wall). Pick up some of the cornstarch-and-oil paste on the wooden spoon. This works best if a few stable dribbles are hanging from the spoon. Slowly move the spoon and its dribbles closer to the balloon.

WHAT WILL I SEE? When the spoon gets close enough, the paste thickens and moves, independently of the spoon, toward the balloon.

WHAT'S GOING ON? When you rub the balloon, the electrical charge you create on its surface is positive. When the tiny particles of starch that make up the cornstarch come near this positive charge, their negatively charged electrons are pulled toward the balloon. Electrical charges act like magnetic poles—a negative charge attracts a positive one and vice versa. The sides of the starch particles facing the balloon are now more negative (and consequently the sides of the starch particles facing away from the balloon are more positive). The positive sides of the starch particles have no balloon or similar charge to attract them and are therefore dragged along for the ride.

In addition, the whole mixture thickens because, although the charges cannot move from one starch particle to another since they are insulated by the oil, they will pack tightly next to one another as the negative sides of the particles nearest to the balloon are attracted to the positive sides facing away from the balloon.

P.S. Cornstarch is an incredibly versatile product. It is widely used for thickening sauces, but its great culinary benefit is that it only works when heat is applied. If you simply add cornstarch to a small amount of cold water, no thickening occurs; however, pour the mixture into a heated soup or casserole and the water and starch molecules begin to bond. The starch particles enlarge, trapping the water as they grow. At about 150°F, the structure of the starch breaks up and forms a mesh of bonded starch and water molecules, preventing free movement of the water molecules and creating a thick sauce. To learn firsthand about these curious properties of cornstarch, check out the "Mixing Madness" experiment in chapter 7.

And if you aren't familiar with the party trick, here's how it works. Rub an inflated plastic balloon on your clothing and then stick it to the wall. The static electricity generated by the motion will suspend it there since the negative electrons collected by the balloon are attracted to the positive areas of the wall.

○ AIM AND POUR

Why does milk dribble down the underside of its carton if you pour it too slowly?

It's not only milk, of course. Many liquids poured from cartons, such as orange juice, result in a sticky floor or ruined shoes. And it's difficult to avoid the effect because when the carton is full, you have little choice but to tip it gently when trying to fill your glass.

WHAT DO I NEED?
+ cartons of milk or other liquids
+ glasses
+ a cloth (for cleaning up afterward)

If you want to demonstrate the effect further, you'll need:

- a vertical cylinder (a liquid dishwashing soap bottle or a wine bottle will do)
- a lighted candle

And, if you are feeling very confident, you'll need:

- a hair dryer
- a table tennis ball

WHAT DO I DO? Open the carton and pour the milk into the glasses. Vary the speed at which you pour so that you range from just a dribble to a rapid stream.

WHAT WILL I SEE? At low pouring speeds, the milk will cling to the edge of the carton and dribble its way down the container before depositing liquid on the floor—this is where the cloth will come in handy. At faster speeds, the liquid will pour freely, allowing you to fill your glass with aplomb.

WHAT'S GOING ON? When the carton of liquid is tipped, the surface of the liquid in the container is raised and moves toward the opening of the container. As the carton is tipped further, liquid pours from the opening, creating pressure at the opening. In addition to this pressure force, there are surface tension forces acting on the fluid that tend to draw it toward the surfaces of the container. At high pouring speeds, the pressure force is much greater than the surface tension forces, and the fluid will leave the carton in an orderly fashion, following a predictable parabolic path to your glass.

However, at low pouring speeds, a point is reached where the surface tension forces are sufficient to divert the path of the fluid jet so that it fails to leave the opening cleanly and becomes attached to the top face of the carton. Once attached, a jet of liquid will tend to stick to that surface thanks to the surface tension forces and a phenomenon known as the Coanda effect. This occurs when a fluid jet on a convex surface (such as water from a tap curving around the back of a spoon) generates internal pressure forces that effectively suck the jet along the surface.

The combination of surface tension and the Coanda effect enable an errant flow of fluid to negotiate the bend from the top face of the carton onto the carton's side and, ultimately and rapidly, onto the floor—or your shoes.

The Coanda effect (also known as "wall attachment") is named after Henri Coanda (1886–1972), who invented a jet aircraft that was to be propelled by two combustion chambers, one on either side of the fuselage and pointing backward. These were positioned near the front of the aircraft. To Coanda's horror, when the engines were ignited, the jets of flame, instead of remaining straight and pointing directly out of the back of the engines, clung to the sides of the fuselage all the way to the tail. While this discovery was obviously troublesome at the time, Coanda has since been immortalized, with the effect now bearing his name.

You must also take into consideration one more effect that causes erratic emission of liquid from cartons. This is the "glugging" that occurs as air is sucked into the carton to replace the lost fluid. This causes the fluid jet to oscillate, leading to intermittent surface attachment and wet shoes even at relatively high pouring speeds.

P.S. The Coanda effect is seen in many circumstances because of the general tendency for fluid flows to wrap around surfaces. Another demonstration that you can try consists of taking a vertical cylinder (this is where the liquid dishwashing soap or wine bottle comes in handy) and placing a lighted candle on the far side away from you. When you blow against the near side of the bottle, the candle is extinguished despite being apparently protected from the breeze. The current of air wraps around the bottle and links up again on the far side rather than being deflected away.

Now take the hair dryer, switch it on to a cool setting, and aim the air stream vertically upward. The effect works best if the hair dryer's nozzle is roughly the same size as the table tennis ball. You'll be able to take the table tennis ball and position it in the air flow above the hair dryer so that it will bob about quite happily without falling (you may need a few attempts to position it because finding the right spot can be difficult). Again, this is a case of the air stream sticking to a surface, this time to the table tennis ball, and the ball is held in place by the Coanda effect. Because this effect is so powerful, you'll have to tilt the hair dryer steeply away from the vertical before gravity can win out.

WANT TO READ MORE? A photograph of the Coanda, the first true jet aircraft, built in 1910, can be found at http://www.allstar.fiu.edu/aero/coanda.htm, alongside information about the designer.

⊙ SINK OR SWIM

Sauce packets in a plastic bottle filled with water can be made to act like submarines. How?

This astounded the *New Scientist* research team when we read about it, so we tried it out in the office and it's true. You can make a sauce packet—the kind you find in fast-food restaurants for serving ketchup, mustard, and mayonnaise—rise and fall in a plastic bottle just by squeezing the bottle.

WHAT DO I NEED?
- a two-liter plastic bottle
- water
- unopened packets of ketchup, mustard, mayonnaise, or any other sauce

WHAT DO I DO? Fill the bottle to the very top with water and add the packets. It helps if you have more than one packet to experiment with because you need to find one (or more) that floats beneath—but just touching—the surface. (Racing different sauces to the bottom of the bottle can become quite addictive.) Screw the bottle top on as tightly as you can. Then squeeze the bottle hard.

WHAT WILL I SEE? The sauce packet will sink to the bottom of the bottle. It's quite amazing. And when you release the pressure on the bottle, it will rise again.

WHAT'S GOING ON? When the bottle is squeezed, pressure is applied to the water. But because liquids are generally very resistant to compression, this pressure is transferred to the packet, which in addition to the liquid sauce contains a small amount of gas. Gas is easily compressed, and as the packet is squeezed by the surrounding water, the space taken up by the packet is reduced. As its volume shrinks, it becomes denser and reaches a point at which it can no longer float. The packet then sinks to the bottom of the bottle, returning to the surface only when the pressure falls—that is, after you stop squeezing the bottle.

P.S. As a floating object falls or rises in liquid, the forces around it change, as does the object's volume. At the surface, it is less compressed (and therefore less dense) than it is at depth. In fact, a suitably compressable object can sit in a liquid at equilibrium neither rising nor falling. Much of this was discovered by Archimedes in the third century BC.

Try squeezing the bottle with differing pressures until you get a packet to hover halfway between the top and the bottom of the bottle. (If you find this tiring, try using a G-clamp or even squeezing the bottle gently between a door and its jamb.) This is essentially how a submarine operates. To control its buoyancy, a submarine has ballast tanks that can be filled with air or water. On the surface, the tanks are filled with air and the submarine's density is less than that of the surrounding ocean. To dive below the surface, the submarine's tanks are filled with water and the air vented until its overall density is greater than the surrounding water. A supply of compressed air (much reduced in volume from normal atmospheric pressure and similar in pressure to the air in the sauce packets after the bottle is squeezed) is kept on board. This is used to force water from the ballast tanks and refill them with air to allow the submarine to resurface. In order for the submarine to operate at precise levels between the surface and the seabed, special trim tanks carry a fine balance of air and water. These carefully adjust the overall density of the submarine to allow it to achieve neutral buoyancy at whatever depth the captain chooses.

9 CITRIC SECRET

Why does lemon juice stop cut apples and pears from browning?

Lemon juice might not improve the flavor of the other fruits, but it certainly works, as many cooks know. And if you want to eat the apples or pears afterward, you can always wash off the lemon juice.

WHAT DO I NEED?
- apples
- pears
- a knife
- two plates
- lemon juice

If you want to take this experiment further, you need:

- some celeriac
- filter paper (you can cut up a paper coffee filter)
- apple juice

- orange juice
- a vitamin C drink

WHAT DO I DO? Slice the apples or pears so they have large areas of their inner flesh exposed to the air. Place half of them on one plate and half on the other. Sprinkle the fruit on one of the plates with lemon juice, and leave the other untouched.

WHAT WILL I SEE? The fruit on the plate that has not been sprinkled with lemon juice will turn brown on its exposed surfaces much faster than the fruit covered in the juice.

WHAT'S GOING ON? To understand why this happens, you first need to understand why some plant tissues turn brown in the first place. Plant cells have various compartments, including ones known as vacuoles and plastids, which are separated from each other by membranes. The vacuoles contain phenolic compounds that are sometimes colored but usually colorless, while other compartments of the cell house enzymes called phenol oxidases.

In a healthy plant cell, membranes separate the phenolics and the oxidases. However, when the cell is damaged—by cutting, for example—phenolics can leak from the vacuoles through the punctured membranes and come into contact with the oxidases. In the presence of oxygen from the surrounding air, these enzymes oxidize the phenolics to produce enzymes such as polyphenol oxidase, which helps to protect the plant and heal the wound—a process also mentioned in "Banana

Armor" in chapter 1. The drawback is that they also turn the plant material brown.

However, this browning reaction can be blocked by one of two agents, both of which are present in lemon juice. The first is vitamin C, a biological antioxidant that gets oxidized instead of the apple's phenolics. The second agent is organic acids, especially citric acid, which make the pH more acidic than the oxidases' optimum level and thus slow the browning. Lemon juice not only has more than fifty times the vitamin C content of apples and pears, it is also much more acidic than apple or pear juice, as a quick taste will tell you. This means lemon juice will immediately prevent browning.

You can also prevent cut apples from browning without lemon juice by putting them in an atmosphere of nitrogen or carbon dioxide, thereby excluding the oxygen required by the oxidases, but this option is clearly not open to the home experimenter or cook.

An excellent vegetable for observing browning is celeriac. Cut a large, relatively uniform slice of this root and then lay several small filter paper discs on the cut surface, each soaked in different solutions with varying levels of vitamin C and citric acid, such as the lemon juice, apple juice, orange juice, and vitamin C drink. If the disc is carrying an agent that blocks the action of oxidases, it will leave a white circle on the otherwise brown surface of the celeriac. Check the differences in browning prevention among the different solutions.

P.S. The pH of a solution is the measure of how acidic or alkaline it is. Most substances fall on a scale of 1 to 14 with 7 being neutral. Distilled water is neither acidic nor alkaline, so it registers a pH of 7, while acids, such as citric acid, fall between 1 and 7, and alkalis, such as sodium bicarbonate, fall between 7 and 14. The more acidic a solution is, the lower its pH, while the more alkaline a solution is, the higher its pH.

Polyphenol oxidase was discovered in mushrooms by Christian Schönbein, a Swiss chemist. It is found in humans, most animals, and many plants. In plants, its function is to protect against insects and microorganisms when the skin of the fruit is damaged. The dark brown surface formed by the skin is not attractive to insects or other animals, and the compounds formed during the browning process have an antibacterial effect.

In some foodstuffs made from plants, this browning effect is desirable. For example, in tea, coffee, or chocolate, it produces their characteristic flavor. However, in other plants and fruits, such as avocados, apples, and pears, browning is a problem for farmers because brown fruit is not acceptable to consumers and, even more important, it doesn't taste as good.

⊃ BUBBLE NUTS

Can you wash your clothes with chestnuts?

Not all detergents come in packets; some grow on trees. Take horse chestnuts, for instance. When you're not roasting them on an open fire or baking them into stuffing, you could be using them to make your whites whiter.

WHAT DO I NEED?

- a handful of chestnuts
- a kitchen knife
- a chopping board
- a pan
- water
- a stove
- a dish towel
- a washing basin
- a bottle with a screw top
- some dirty socks

WHAT DO I DO? Remove the brown outer casing of the chestnuts, chop them up into small pieces, and put them in the pan. Add a cup or two of water and boil the chestnuts for a few minutes, then let them cool. Strain the mix through the dish towel into the basin to remove the solids while keeping the liquid. Pour this liquid into a bottle and shake it. Now put it back in the basin and wash your socks in it.

WHAT WILL I SEE? You should see a soapy lather form on the liquid you pour into the basin. And, if you give your socks a good scrub, you'll see how clean they emerge.

WHAT'S GOING ON? This highlights a property of chestnuts that few people are aware of. Horse chestnuts contain saponin, a natural soap or surfactant. Many other plants, including soapwort, produce saponins, which help protect them against disease because they are toxic to bacteria and fungi. As you have seen, they can be extracted with water—a trick that has been used for centuries to make a soapy liquid for cleaning linen.

Surfactant molecules have a polar region that is attracted to water molecules (it is hydrophilic) and a nonpolar region that is repelled by water (hydrophobic). They can therefore be dissolved in water as well as other kinds of organic solvents, including the substances that make your socks dirty. It also makes them very similar to synthetic detergents and very good for cleaning.

Because saponins are very mild surfactants, they are less damaging to materials than stronger, synthetic detergents. This is why they are popular with art conservators, who use them to clean delicate fabrics or ancient manuscripts.

Surfactants also play a role in firefighting. They are, along with a number of chemicals, a constituent of Light Water, a substance produced by the company 3M. As a dilute solution in water, it forms a light, stable foam used primarily to fight fires that involve flammable organic liquids such as oil. The foam floats on the surface of the burning liquid (hence its name), covering it with a thin surfactant film that prevents further evaporation of the liquid and extinguishes the fire by cutting off the oxygen supply.

P.S. You need to be aware that the horse chestnut extract you have created is mildly poisonous, causing coughing and sneezing if drunk, and that it can also be a skin irritant (you may want to use kitchen gloves while you do your laundry). Close adult supervision of children is required. And when you've finished, wash all the utensils you used to make sure there are no traces of saponin left.

◐ HOT CHOCOLATE

Can you measure the speed of light using nothing more than a chocolate bar and a microwave oven?

Yes, although a counter theory says that all you need is a chocolate bar and a classroom of schoolchildren just waiting to pounce. . . . This is a marvelous experiment that actually allows you to measure one of the fundamentals of science—the speed of light—in your own home.

WHAT DO I NEED?
♦ a bar of chocolate (the longer the better)
♦ a metric ruler
♦ a microwave oven

WHAT DO I DO? Remove the turntable mechanism from your microwave oven—the bar of chocolate needs to be stationary. Put the chocolate on a plate in the microwave and cook at a high setting until it starts to melt in two or three spots—this usually takes about forty seconds. You should stop after sixty seconds maximum, for safety.

WHAT WILL I SEE? Because the chocolate is not rotating, the microwaves are not evenly distributed throughout the bar, and spots of chocolate will begin to melt in the high-intensity or "hotspot" areas. Remove the bar from the microwave and measure the distance in centimeters between adjacent globs of melted chocolate.

WHAT'S GOING ON? The frequency of the microwaves is the key here. A standard oven will probably have a frequency of 2.45 gigahertz (GHz). If your oven is 2.45 GHz, this means the microwaves oscillate 2,450,000,000 times a second. You will need to adjust this figure depending on your oven's frequency, which should be printed on the back of the oven or in its instruction manual. (If the frequency is given in megahertz [MHz], you simply need to know that 1,000 MHz is equal to 1 GHz.) Microwaves are a form of electromagnetic radiation and therefore travel at the speed of light. Once you know the frequency of the microwaves in your oven, finding out their wavelength will allow you to calculate how fast they are traveling.

This is where the chocolate comes in. The distance between the globs of molten chocolate is half the wavelength of the microwaves in your oven, so double the measurement you have taken of the gap between the molten globs to find the microwaves' wavelength. In the *New Scientist* microwave oven, the distance between the globs of molten chocolate was six centimeters, so the wavelength in our 2.45 GHz oven is twelve centimeters.

To calculate the speed of light in centimeters per second, you need to multiply this wavelength by the frequency of the microwaves: $12 \times 2,450,000,000 =$

29,400,000,000, which is astoundingly near to the true speed of light of 29,979,245,800 centimeters per second (or 299,792,458 meters per second, as it is usually expressed). Since scientists have universally adopted the metric system for measurements, the speed of light is defined in meters per second.

Try it yourself, measuring as accurately as possible to get a figure even nearer to the true speed. If your chocolate bar is chilled beforehand, the molten areas tend to be more distinct when they first appear. Of course, you may find testing a variety of different chocolate bars, all of which taste delicious slightly melted, as an aid in your research. Dedicated scientists know that they must always double-check results.

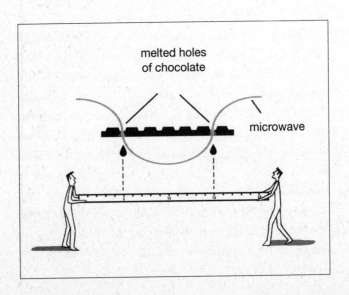

melted holes
of chocolate

microwave

❾ SUCKING EGGS?

How can you force an egg into a bottle?

This classic experiment has been repeated many times through the years, but the forces at play are often misunderstood. In fact, the illuminating part of this experiment isn't getting the egg into the bottle, but getting it out. . . .

WHAT DO I NEED?
- a hard-boiled egg
- a glass bottle with an opening slightly smaller than the circumference of the egg when it is placed upright atop the bottle (milk bottles are ideal; wine bottle necks tend to be too narrow)
- long matches

Vinegar and syrup will allow you to carry out a second experiment.

WHAT DO I DO? Shell the egg so you have the fresh, rubbery white exposed. Place the bottle on a flat, flame-resistant surface. Light a match and drop it into the bottle, making sure it stays lit. Add a few more burning matches as quickly and as safely as possible—if you can, add all the matches at once. Put the egg, pointed side down, over the neck of the bottle.

WHAT WILL I SEE? The matches will soon go out, because they are starved of oxygen once the egg seals the bottle, and the egg will be forced into the bottle, looking rather peculiar as it contracts in order to slide down the inside of the neck.

WHAT'S GOING ON? Inside the bottle, the lighted matches heat up the air, causing it to expand. This means that some of the expanded air escapes from the bottle. When the matches go out after the egg has been placed on top of the neck, the air contracts, creating a lower pressure inside the bottle than exists outside. The pressure outside the bottle forces the egg inside, and it plops to the bottom.

While it seems as though the egg is being pulled or sucked into the bottle, in reality it is pushed by the outside pressure. The greater pressure outside forces the egg down into the lower-pressure area in the bottle.

While many people were taught that the egg is sucked into the bottle when the burning matches use up the oxygen in the air inside the bottle, this is just plain wrong. It's the change in the temperature of the air after the matches go out that causes the air in the bottle to contract. An egg works well because its moist, rubbery surface forms an effective seal around the neck of the bottle. If you repeat this experiment with a chocolate egg or an unpeeled, uncooked new potato that doesn't completely seal the neck of the bottle, air will be able to return to the bottle and equalize the pressure between the inside and outside without carrying the object with it.

You can confirm this when it's time to get the egg out of the bottle. Hold the bottle on its side and carefully jiggle the egg so that its narrow end is resting against the bottle's neck. Turn the bottle upside down and form a seal between your mouth and the bottle opening. Now blow hard, which will increase the air pressure inside the bottle and force the egg out. But be careful: the egg can come out at quite a speed!

P.S. While you have your eggs out, you can try another experiment, especially if you've saved some vinegar after trying out the "Plastic Milk" experiment on page 1.

Place a whole uncooked egg in a jar of vinegar and leave it for a couple of days. When you return you'll see the shell has disappeared and the egg has swollen to perhaps twice its size. This is because the shell, which is made of calcium carbonate, reacts with the vinegar's acetic acid to create carbon dioxide, water, and calcium ions. The water passes into the egg through its now exposed membrane, driven by the pressure of osmosis (the passage of water from a region of high water concentration through a semipermeable membrane to a region of low water concentration, reducing the difference in concentrations). The white of the egg has a very high concentration of protein, and water passes from the vinegar into the egg in an attempt to equalize the concentrations on either side of the membrane.

To reverse this process, place the egg in a solution of 75 percent sugar syrup and 25 percent water. After a couple of days the egg will shrink to less than its original size as osmosis works in the opposite direction, and water leaves the egg to equalize the concentration between the egg and the sugary syrup mix surrounding it.

2 IN THE DINING NOOK

9 SAUCY STUFF

What's the best way to get tomato ketchup out of a bottle without breaking the bottle or splattering the sauce?

Some would insist, of course, that getting ketchup out of a bottle is an art, not a science. We at *New Scientist* disagree for, as you will discover, there seems to be a lot of physics involved. The arrival of squeezy bottles has diminished the fun. For purists, the glass ketchup bottle remains the ultimate challenge.

WHAT DO I NEED?
- a meal requiring tomato ketchup (it's not essential, as you can do this experiment using an empty plate, but there's no doubt that French fries enhance the experience)
- a glass bottle of tomato ketchup

WHAT DO I DO? Attempt to shake the ketchup out of the bottle onto your plate without splattering it or breaking the bottle. Ideally, you'll be able to release the sauce in pleasing globs that form an ordered pile in one part of the plate.

WHAT WILL I SEE? It depends on your skill. If you follow one of the methods described below, you may be successful in getting all the ketchup to land on your plate. Technique and practice seem to be the keys.

WHAT'S GOING ON? Science is, of course, involved in extracting the ketchup from the bottle. The worst thing you can do, as the first of our methods demonstrates, is to bang the bottom of the bottle to get the ketchup moving. While you will see this technique employed widely in diners and hamburger joints, it is of little use. Instead, you should turn yourself into a centrifuge or exploit the thixotropic nature of ketchup.

Here I present a few of the best methods supplied by *New Scientist*'s readers, each named after the writer. You should try them all before settling on the one that suits you best. All, as you would expect, employ sound scientific principles, and all have their champions among the magazine's staff.

The Foy (or inertia) method: Most people hit the bottom of the upturned bottle, which pushes the bottle away from them. Because of the inertia of the ketchup, this only ensures that if it moves in the bottle, it will flow in the opposite direction to the one in which you want it to go. The sauce is pushed further toward the bottom of the bottle, rather than out of it. To get the last drop of ketchup out, hold the upturned bottle over your plate and hit the underside of the wrist of the hand holding the bottle with your other fist, jerking the bottle upward. The inertia of the sauce will now eject it from the bottle.

The Wong (or centrifuge) method: First, put the lid on the bottle and grip it at its base. Then swing your arm as if you were throwing a ball overhanded. This method, which uses the principle behind a centrifuge, forces the ketchup to the top of the bottle, allowing you to pour it out. (Whether you can use such a flamboyant technique in a fancy steak house is open to question.)

The Lloyd-Evans (or thixotropic) method: Ketchup is gloopy because it is thixotropic. This means that, when it is at rest, it has a thick gelatinous consistency that can be altered to a runny consistency by the input of energy, typically by shaking. (This is in contrast to dilatant materials, as you'll see in the "Mixing Madness" experiment in chapter 7.) The thixotropy is provided by the starch used in ketchup. Starch molecules come in the form of long chains. When starch powder is mixed with water and heated or subjected to enzyme treatment, weak links are formed between these long molecules. When the ketchup is made at the factory, there is thus a physical change in the sauce, creating a pasty, gelatinous matrix. The thickening and gelling abilities of starches such as corn flour, rice flour, potato flour, and powdered arrowroot are used in sauces, gravies, and soups, as seen in the "Shape Shifting" experiment in chapter 1. To get the ketchup out of the bottle, first ensure that the lid is on (unless you want to upset the person sitting directly opposite you), then give the bottle some vigorous but not overathletic shaking. This will break some of those weak bonds between the starch molecules. Now turn the bottle upside down over your plate and watch the ketchup emerge in a slow, gentle stream.

The Bellis (or vibration) method: Rapidly hit the side of the bottle with the soft edge of your fist, from your little finger to your wrist, while holding the bottle at an appropriate angle. The vibration of the rapid thumping will break down the structure as above, allowing the ketchup to flow easily. This is the same principle used to settle concrete into molds. In the case of concrete, however, a vibrating device is probed into the mix and the rapid vibrations shake the concrete, allowing it to spread down into the mold.

The Hann (or temperature) method: To get a nice even coating on your fries, the best thing you can do is to warm up the ketchup. Provided you remove the metal cap, a burst of *no more* than fifteen seconds in a microwave usually does the trick. Obviously, the amount of sauce in the bottle and how chilled it was will affect the success of this method, but heat is a guaranteed way of making the ketchup runnier as the bonds between the layers of molecules will break under sheer stress and the molecules will slide past one another more easily. This property of viscous fluids—the phenomenon by which their viscosity tends to fall (or, alternatively, their fluidity increase) as their temperatures increase—is known as the temperature dependency of liquid viscosity.

The Goldstein (or poking) method: Take a long, thin object (a chopstick works well) and poke a hole into the ketchup through the mouth of the bottle. This breaks the air/liquid seal and lets the sauce flow freely.

The Medhurst (or "what-not-to-do") method: Leave the bottle in the back of your kitchen cupboard for a few years until fermentation sets in. Pressure will build up inside the bottle so that when the lid is removed the sauce will eject dramatically from the opening. This certainly removes the ketchup from the bottle, but whether it fulfills the "no splattering" criterion is open to debate. Howard Medhurst, who suggested this method, admits that his kitchen now has a four-inch-[10 cm-]wide red streak running across the entire ceiling, one wall, and half of the floor.

P.S. Ogden Nash, renowned for his witty, light verse, had something to say on the subject. His granddaughter, Frances R. Smith, of Baltimore, Maryland, who is the family representative of Nash's back catalog, kindly supplied us with the following poem, *The Catsup Bottle:*

> First a little
> Then a lottle
> The catsup bottle.

9 TUNEFUL O' JOE

Why does the note coming from a spoon repeatedly hitting the inside of a newly poured cup of instant coffee rise in pitch?

This is the kind of thing that you may have always been subconsciously aware of, yet never seen put into words.

But now that you know about it, you'll just have to try it. So make yourself a cup o' joe and get stirring.

WHAT DO I NEED?
+ a large ceramic mug
+ instant coffee or powdered hot chocolate
+ boiling water
+ a teaspoon

WHAT DO I DO? Put the instant coffee or hot chocolate into the mug, add the boiling water, and stir thoroughly. Hold your spoon vertically in the liquid and listen carefully as you repeatedly tap the base of the mug for a minute or so.

WHAT WILL I HEAR? The sound of the spoon hitting the bottom of the mug gradually rises in pitch.

WHAT'S GOING ON? The reason why this happens has been a matter of debate for some time.

When the mug is tapped, it acts somewhat like a bell: the sound generated is determined mainly by the material of the mug and its thickness, diameter, and height. It is also determined by the distance between the surface of the water and the bottom of the mug, which causes the mug to resonate at a particular frequency. It is tempting, therefore, to speculate that stirring the water, which causes it to sink slightly in the middle and rise a bit along the sides of the mug, is responsible for the change in pitch. But this slight change in water level is not nearly enough to account for the big change in pitch, which can sometimes exceed a full octave. Also, the pitch change occurs even when the water is not rotating, so something else must be going on.

And that something is air. When you pour hot water on instant coffee, many of the volatile components in the coffee form tiny bubbles. Additionally, water frequently has air dissolved in it, so when it is added to powders such as coffee, nucleation sites form around the grains and more bubbles are created (as discussed in the "Bouncing Rice" experiment in chapter 1). The air bubbles lower the speed of sound in the water by temporarily making the water more compressible (less "springy"). It takes longer for the sound wave to travel between the bottom and the surface, and thus the pitch falls when the instant coffee is first added. Then, as the bubbles rise to the surface and burst, the sound travels faster and the pitch rises.

The greatest difference in pitch occurs when the spoon is tapping the bottom of the mug. This is because the sound waves have further to travel from the base of the mug to the surface of the liquid than they do from the sides.

P.S. Because seemingly thousands of cups of coffee or tea are consumed in the *New Scientist* office each week, staff were asked to experiment with their daily cups. They found the effect in cups of tea was tiny however much milk was added or however vigorous the stirring. But in instant coffee or hot chocolate, there was a clear result.

Over a period of about five seconds after the boiling water was added to the powder, the note of the spoon against the mug rose distinctly as the drink was stirred.

Bubbles could be seen rising to the surface, which suggests that they were the key factor in the change of sound. Tea produces fewer bubbles than instant coffee or hot chocolate, which explains why the effect was more difficult to reproduce in tea.

WANT TO READ MORE? Frank S. Crawford's paper "The Hot Chocolate Effect" was published in the *American Journal of Physics* (volume 50, number 5, May 1982, page 398).

9 SILVER LINING

If you hold an egg in a candle flame until it becomes sooty and then dunk it in water it looks silver. Why?

This old but extraordinary experiment allows you to use a hopeless pun as you point out that "every egg has a silver lining." What's more, because it's another question involving eggs, you can try it out alongside the "Green Eggs and Cabbage" and "Sucking Eggs?" experiments in chapter 1.

WHAT DO I NEED?
- a white or pale brown chicken egg, uncooked
- matches
- a candle
- water
- a glass bowl

WHAT DO I DO? Light the candle and hold the egg between your index finger and thumb. Because you are going to move the egg into the flame, this is an experiment children will have to watch from the sidelines. Hold the egg near the top of the candle flame so that it begins to become covered in soot. Rotate it so that as much of the egg as possible becomes sooty. Fill the bowl with enough water to cover the egg and carefully place the egg in the water.

WHAT WILL I SEE? It's quite astounding. Once in water the egg looks neither black nor egg-colored, but silver with an almost mirrorlike quality.

WHAT'S GOING ON? The silver appearance of the egg's surface is created by a very thin layer of air that gets trapped between the egg and the body of the fluid, because the soot-covered surface repels liquid. The air forms a film around the egg. Light passing through the water strikes this layer of bubbles and is reflected back, like a mirror. Eventually, the air bubbles will dissolve into the water and the shell will appear sooty again.

P.S. The good news is that if you use the eggs immediately after carrying out this experiment, they are still edible. If you want to be truly ecologically sound, you could even use the water from the bowl to boil them.

◯ TEETHING TROUBLE

Why does eating spinach make your teeth feel weird?

Sadly, this works better with canned spinach than with fresh spinach, but it is still one way to get your kids to eat their green vegetables. But be cunning: make sure that they think they are doing the experiment rather than being experimented on.

WHAT DO I NEED?

- plenty of spinach leaves (preferably canned, heat-processed spinach)
- a pot of boiling water
- a colander
- a plate
- a fork

WHAT DO I DO? Boil the spinach until it is cooked, drain it, let it cool a little, and eat it. Then run your tongue around your teeth and mouth.

WHAT WILL I FEEL? Your teeth and the inside of your mouth will feel fuzzy and furry. This may be one of the reasons why children particularly dislike spinach, but once they realize they are part of a noble experiment, their attitudes change radically.

WHAT'S GOING ON? Spinach contains a large amount of oxalate crystals—mineral salts of oxalic acid. When spinach—especially the canned variety—is cooked, some of the spinach cell wall structure is damaged and the oxalate crystals leak out. These coat your mouth to give it the fuzzy feeling. Fresh, uncooked spinach does not produce a similar effect because its cell wall structure has not been damaged. Spinach is also rich in calcium and oxalic acid, and these combine with the calcium in saliva to deposit large amounts of furry, calcium-rich plaque on your teeth.

P.S. Chard and beet leaves have a similar effect.

WARNING: People who have problems with renal function should avoid spinach and other oxalate-rich foods because of the increased risk of producing oxalate stones in their kidneys.

⟲ METAL ANGUISH

What causes the pain induced by a piece of aluminum foil on a tooth filling?

If you have looked after your teeth well, you can smugly sit this one out. Those with fillings, however, will quickly become enlightened.

WHAT DO I NEED?
- aluminum foil
- saliva
- a tooth with a filling (still in situ)

WHAT DO I DO? Generate a decent amount of saliva in your mouth. Pop the foil in and position it over a filling in your teeth, preferably a molar so you can bite on the foil.

WHAT WILL I FEEL? You may jump! You'll feel anything from a slight tingling in the filled tooth to actual pain as the foil touches it.

WHAT'S GOING ON? When two dissimilar metals are separated by a conducting liquid, a current will flow between them, and this can stimulate nerves. In this case the two dissimilar metals are the amalgam in your tooth and the aluminum in the foil. A thin film of saliva separates the foil from the filling and, because it is a reasonable electrolyte containing various salts, it acts as the conducting liquid, allowing a current to flow between tooth and filling. Because the filling is close to the dental nerve, the current will stimulate it, causing pain.

P.S. Luigi Galvani first discovered the dissimilar metal/electrolyte effect in 1762, when he carried out experiments using the nerves in frogs' legs. When probes of differing metallic compositions were applied to the frogs, their legs twitched. Using tooth amalgam and spit is a more humane way of carrying out this experiment, although the owner of the filling might not agree.

○ FLOATERS

Why do floating pieces of breakfast cereal stick to each other and to the sides of the bowl?

While it's been suggested that Cheerios and crisped rice cereals such as Rice Krispies herd together as a defensive strategy to avoid being eaten, this clearly is not successful. Other forces are at work. . . .

WHAT DO I NEED?
- a bowl of milk
- a wheat- or rice-based breakfast cereal

- three polystyrene cups
- tap water

WHAT DO I DO? First check out the basis of the question. Pour a bowl of milk and add the breakfast cereal of your choice. Don't overfill the bowl; just add enough to cover a third of the surface at most. Then work out the physics involved using the polystyrene cups.

Fill the first cup with water to within half an inch of the rim. Fill the second cup to the top and then carefully add more water until the meniscus is above the top of the cup but not spilling over—in other words, the water is held in a convex bulge above the top of the cup by surface tension.

Now place a half-inch-diameter circle of polystyrene torn from the third cup in the middle of each of the first two.

WHAT WILL I SEE? In the first experiment, the breakfast cereal will clump together, preferring the edges of the bowl to the center. You can now eat the cereal.

In the second experiment, the piece of polystyrene floating in the partially filled cup will, with a little prompting, move to the side of the cup and be held there. By contrast, the piece floating on the convex bulge of the water in the second cup will remain near the center. Furthermore, if you push the piece to the edge of the cup with the tip of a pencil, the edge will repel the small piece, driving it toward the center with considerable force.

WHAT'S GOING ON? The effect is caused by the surface tension of the water. In the partially filled cup, the water surface curves up to meet the polystyrene wall because water molecules are more attracted to polystyrene than to one another.

The water forms the convex bulge at the top of the full second cup because the surface tension constrains the liquid surface to the smallest area possible, which similarly accounts for the spherical shape of liquid drops. The water also curves up to meet the small circle of polystyrene on all sides. Where the water meets the polystyrene of the small circle, the surface tension pulls on each contact point in the downward and outward direction determined by the angle of contact with the water. When the circle is in the middle of the cup, the pull on the circle on one side is directly balanced by the pull on the opposite side, because the water curves up to meet the circle equally at all points. This means the polystyrene circle is held in the middle of the liquid in the cup.

However, when the piece is moved toward the side of the partially filled cup, the upward curve of the water surface at the side of the cup reduces the curve of the surface in contact with the circle as these two curves come together. This increases the outward pull on the side of the circle nearest to the cup edge, resulting in a net force toward the side of the cup. As a result, if the cup is only partially full, the polystyrene circle will move away from its central position and sit at the side, unlike the polystyrene circle in the full cup.

The effect accounts for the clumping of those cereal

pieces on the surface of milk in your bowl and also for the similar behavior of leaves and twigs on ponds and lakes.

◯ APPLE FOOL

Is it true that you can fool yourself into believing you are eating something other than the food that's in your mouth?

It's certainly true that the senses can be fooled. I have personal experience of how one sense can be perceived as another. When a virus temporarily took away my sense of smell, I was astounded by how easy it was to confuse one sensation with another. I'd often leave an air-conditioned office on a freezing day, sense the chilly air, and think, Wow! My sense of smell has returned. Somehow, my brain was confusing the sense of touch, which detected the cold, with the return of my sense of smell. This experiment shows how easily human sensations can be fooled. And, after fooling yourself, you can read about synesthesia.

WHAT DO I NEED?
◆ a sliced pear
◆ your nose
◆ a sliced apple
◆ your mouth

WHAT DO I DO? Hold a piece of the pear under your nose while eating a slice of the apple.

WHAT WILL I TASTE? Despite the fact that you are eating an apple, you'll think you are eating a pear.

WHAT'S GOING ON? The confusion arises in part because most people think that they taste using their taste buds. We don't. We all detect flavors via our sense of smell.

Indeed, most of what we call taste is actually flavor produced by the smell of food passing from our mouths into our nasal cavities, where we detect it through olfaction. True taste is only the bitter, sweet, salt, sour, and umami (savory) detected by the taste buds. Bite into a strawberry and your tongue only tells you that it is sweet, just as it would if it were chocolate. It's the odors rising through your throat to your nose that tell you that that particular sweetness is strawberry-flavored.

If you remove the sense of smell or block those odors, or—as in the experiment above—replace them with something else, you can confuse your senses of taste and smell. Your taste buds detect the apple in your mouth as being sweetly acidic using its sweet- and sour-detecting taste buds, and your sense of touch recognizes the texture to be that of a type of tree fruit. However, because you can more strongly detect the smell of pear under your nose than you can the smell of apple rising through the back of your mouth, your brain is fooled into thinking that what's in your mouth is a piece of pear.

The experiment works this way around because pears have a stronger smell than apples.

P.S. People who have synesthesia make bizarre sense associations all the time. No one is really certain, but this condition is thought to involve crossed wires in the brain. We are born with many connections to different areas of our brain, but as we mature our brain cuts back on those it doesn't need. Synesthetic connections are possibly some of those that were not pruned during infancy. The extraordinary result is that some people—as many as one in twenty-five—perceive things like letters, words, numbers, or days of the week as having very distinct and immutable colors. A smaller group even experiences sounds as having colors, or flavors as having shapes. One synesthetic person reported that she knew when the milk was about to go bad because its shape began to quiver.

○ STIRRING STUFF

If you are stirring a drink such as tea with loose leaves, why do the leaves head to the middle of the teacup the faster you stir?

Intuition suggests that objects should fly outward not inward, so what's going on? There's no truth in the rumor that tea leaves possess supernatural qualities as some fairground entertainers would have you believe, so we have to fall back on that old stalwart of truth—science.

WHAT DO I NEED?
* loose-leaf tea (or a tea bag, cut open)
* a cup or mug
* hot water
* a spoon

WHAT DO I DO? Place the leaves in the mug, add water, stir the drink, and watch what happens. If you feel thirsty, you can always drink the tea later, but if you are used to drinking only the product of tea bags, be prepared to strain the leaves through your teeth.

WHAT WILL I SEE? Any floating tea leaves will move to the center of the mug as you begin to stir. As you increase the speed of stirring, they'll move more rapidly to the center.

WHAT'S GOING ON? The answer lies in what is known as the pressure-momentum balance in the rotating fluid. If the rotating fluid created by your stirring is to remain in the mug rather than breaking over its lip and running down the sides of the vessel, the inertial force created by the fluid rotation must be balanced by a pressure gradient within the liquid. Pressure is lowest in the center and increases toward the wall of the mug. Think of a weight being swung around at the end a piece of string: tension in the string prevents the weight from flying off. This tension is analogous to the pressure gradient in the tea.

When you stir the liquid, it is pushed outward, raising the pressure at the edge of the mug and lowering it at the center. You can clearly see that as the liquid rotates it is forced outward, creating a dimple in the center of the liquid surface. In fact, if it wasn't for the sides of your mug constraining the liquid, it would simply flow out over the tabletop. So any particle that is less dense than the surrounding liquid will tend to move inward to the center of the mug, where the pressure is lowest. This makes sense, but even objects that are heavy move inward. Try dropping a handful of peanuts into the liquid, or even small stones (don't use your best china for this experiment), and you'll see that they still cluster toward the middle when you stir the liquid.

To understand why even heavy objects move to the middle, you have to consider another physical phenomenon inside your mug: friction. Compared to other areas in the mug, the bottom of the mug is rougher and therefore induces more friction against the liquid as it rotates parallel to the surface. This additional friction causes the liquid adjacent to the bottom of the mug to rotate more slowly than the liquid at the surface. This leads to the formation of a circulation current in the mug—liquid moves out across the top surface thanks to the pressure gradient of the stirring motion, then downward nearer the outer wall, inward along the bottom, and back up the mug's center. Granules such as tea leaves or small stones are carried along with this current until they are swept to the center of the bottom of the mug, where the pressure gradient is at its lowest thanks to the friction. Gravity also plays a part. The heavy particles are held at the bottom by a combination of the low pressure and the gravitational pull of Earth. The liquid, which cannot accumulate in a pile at the bottom of the mug, keeps on moving, circulating up the center of the mug along the axis of rotation and away from the bottom.

As the mug is stirred, liquid moves outward across the top surface, down the outer walls of the mug to the bottom, and up the center of the mug to the surface again.

high friction base of mug

route of liquid

Particles carried by the liquid are deposited in the low-pressure area in the center of the mug.

Physicists call this accumulation of particles a boundary layer effect, and the rough bottom of the mug is the key to inducing this property. If you can get hold of smoother vessels with lower friction than the average mug, you'll see the effect is far less marked.

P.S. The forces at work in your mug are the same ones that cause rubbish to collect in street whirlwinds on blustery days rather than being distributed evenly over the road.

3 IN THE FAMILY ROOM

○ FLOWER POWER

How do flowers "drink" water?

If you forget to tend to flowers in a vase, you'll discover the astonishing rate at which they take up water. A full vase can become empty in no time at all, leaving you with drooping stems and fallen petals. So where does all that water go?

WHAT DO I NEED?
- white flowers, such as lilies or carnations
- food coloring (red or blue give the best effect)
- water
- scissors or a knife
- a clear glass vase
- a sunny window ledge

You can also use celery if you want to study this effect further.

WHAT DO I DO? Trim each flower stem with the scissors or knife and place them in a vase containing water plus food coloring. Make sure the color in the vase is distinct—the darker the better. Place the vase in a sunny spot. Check the flowers every hour.

WHAT WILL I SEE? Over the course of a few hours, the petals will change color. The patterns that form can be quite splendid, with veinlike structures across the petal surfaces, and petal edges that are deeply colored.

WHAT'S GOING ON? Plants need water in order to live and grow, and they get it by drawing up water from the soil through their roots (or, in this case, from the water in the vase). The process is driven by transpiration, which is the evaporation of water from petal and leaf surfaces (the French word for "to sweat" is *transpire*). When the leaf surfaces become deficient in water, more is drawn from the vase as water molecules are pulled upward to replace those that have been transpired. Obviously, a sunny, warm spot speeds up this process. Normally, all this is invisible, but the food coloring demonstrates exactly what is going on.

The colored water is drawn through vascular structures (the xylem) in the plant. At its simplest level, the xylem acts like a straw. As water is transpired away from the petal or leaf surface, more water is pulled into the straw to replace it. The xylem conducts water and dissolved nutrients up from the root. You can see the structure of xylem if you remove one of the flowers from the vase and cut through the stem. Veinlike tubes will be clearly visible in the cross-section, filled with colored liquid.

food coloring

Slice through a celery stalk to give
a newly cut surface and place the
stick in the vase containing the
colored water.

After a few hours, remove it and
slice through the stalk again to
reveal the xylem structure.

Over the course of just ten hours, you'll see the colored water spread throughout your flowers, until they are entirely dyed. Plants, especially in hot conditions, can take up vast amounts of water, which is why a vase that was full one morning can be almost empty the next. On a hot day, a full-grown tree, such as birch or sycamore, can draw upward of five hundred quarts from the ground, in some cases aided by root pressure, forcing water upward into the xylem; all of which is later transpired through evaporation.

P.S. Celery stalks show the xylem structure even better than the thin stems of flowers. Remove a celery stalk from a new bunch, slice off the bottom end to expose a newly cut surface, and place the stick in the vase containing the colored water. Leave it for a few hours before returning and removing the stalk. Wipe it clean and cut through the stalk—you'll see distinct dots on the cut surface indicating where the xylem has taken up the colored water.

9   SHOCKING TV

Why, when you switch off your television in a darkened room and touch the screen, is there a crackle of sound and a fluorescent glow at the place you touch?

This is more fun than yet another episode of reality TV and hints at what's really inside your TV—other than Elvis, of course. You'll also feel the hairs on your arm stand up (again, nothing to do with Elvis).

WHAT DO I NEED?
- a non-flat-screen, old-fashioned TV (hopefully you kept yours in the attic or have given it to the kids)
- a darkened room
- a finger

WHAT DO I DO? Switch off the lights, turn off the TV, and touch the screen—don't wait too long or you won't see the effect.

WHAT WILL I SEE? You'll see a fluorescent glow at the point at which you touch the otherwise blank screen and hear a crackle of static electricity discharge. Move your finger or hand across the screen and you'll hear more static and the hairs on your arm will simultaneously become charged in the electrical field of the TV.

WHAT'S GOING ON? An old-fashioned TV has a cathode-ray tube, which creates a picture by passing electricity from the cathode (the negative electrode by which electrons enter an electrical device) at the rear of the TV's vacuum tube to the anode (the positive electrode) at the front. (Modern liquid crystal display screens create the TV picture using a different process.) The anode on an old-style cathode-ray tube is created by a thin layer of aluminum that coats the back of the TV screen. Pictures are created by fluorescence acting on the phosphor, the synthetic fluorescent that also coats this screen. When the TV is switched on, this anode induces a charge on the outside of the screen. When the TV is switched off, most of this charge leaks away, but some lingers on the phosphor. This charge flows when it is touched by your finger and the phosphor glows in the way it would if the screen was still switched on, while the crackle is the sound of the electric discharge.

9 SHEDDING LIGHT

Why does the flame on a candle that is standing on a revolving turntable point inward and not outward?

When we first attempted to replicate this phenomenon in the *New Scientist* office, we couldn't get it to work. When the candle was lit and the turntable switched on, the flame trailed behind the candle as it orbited the center of the turntable in the way it does when you walk along holding a candle. Then somebody suggested using a jam jar. . . .

WHAT DO I NEED?
- a candle
- a turntable (a record player turntable set to 45 rpm—or 78 rpm if you have a very old gramophone—will do, although a potter's wheel or lazy Susan works just as well)
- a large jar

If you want to take the experiment further, you'll need:
- a car
- a helium balloon
- a bubble level

WHAT DO I DO? Attach the candle firmly to the turntable. A heavy base of plasticine helps, but a wide-based,

heavy candle holder is even better. Light the candle. Hold
the upturned jar over the candle and switch on the turntable. Make sure that air can get under the base of the
jar, or the candle will go out when it becomes starved of
oxygen. (You might want to combine this experiment
with the next one, "Air Space,'" because much of the
equipment is the same.)

WHAT WILL I SEE? The candle flame will not be thrown
outward by the motion of the revolving turntable, like
most objects subjected to such rotation, nor will it trail
behind the candle, thanks to the jar protecting it from any
breeze. It will actually point in toward the center of the
turntable.

WHAT'S GOING ON? In effect, the air in the rotating jar is
the object being spun in a centrifuge. This means
the candle flame bends toward the inside of the turntable
for the same reason that it normally points straight
upward: the heated gas that comprises the flame is
less dense than the cooler surrounding air. So as centripetal force (the force that acts on a body moving in a
circular path and is directed toward the center around
which the body is moving) acts on the rotating system, it
is the denser, cooler air that moves outward and the
heated, less dense gas of the flame that moves inward.

We tend to think that the flame is composed of
"something" because we can see it, and that the surrounding air space is composed of "nothing" because we
can't see it. Therefore the brain intuitively suspects that
the flame will move outward as the turntable rotates,
in the way that a piece of cotton or tissue paper would.

But cotton and tissue paper are denser than the surrounding air, whereas a flame is less dense. Therefore the visible flame moves inward while the invisible air moves outward, in apparent contradiction of our brain's interpretation of visible objects.

Another, perhaps simpler, way of understanding why the candle flame points inward is to consider a similar problem. If you are driving a car that contains a helium balloon held by a string and you brake hard, you will slam forward against the seat belt but the balloon will move to the back of the car. This is because the air in the car has inertia and continues forward just as you do, while the balloon reacts by floating toward the area of lowest pressure—the lowest density portion of the air mass, which is found at the back of the car. Similarly, the tethered balloon will lean forward under acceleration and toward the inside of bends as the car turns, in the way the candle flame leans inward as it rotates.

The candle flame, like the balloon, is buoyant, its shape resulting from a complex interaction between the burning wax at the wick and the heating of the surrounding air. So, like the balloon, the flame also floats in the direction of lowest pressure—toward the axis of rotation at the center of the turntable. To complete the comparison, the candle, like the car, is accelerating with respect to the air surrounding the flame, so the air is moving radially outward relative to the candle. The flame reacts by floating inward.

If you do attempt this further extension of the experiment, it should go without saying that the driver should keep his or her eyes on the road and not attempt heavy braking if the conditions are slippery, and the experiment should be carried out by a passenger who will not distract the driver. It should be tried only on private land where there are no other vehicles or pedestrians, and all occupants of the vehicle must wear correctly fitted seat belts.

As a final visual test of this phenomenon, place a bubble level on your turntable, pointing away from the center like a bicycle wheel spoke, and switch the turntable on. The bubble moves inward, not outward, because the liquid inside the instrument is denser than the air bubble, and consequently the bubble is pushed inward.

P.S. Sue Ann Bowling of the University of Alaska tells *New Scientist* that if she were being really picky she would argue that the less dense candle flame is accelerated more by the centripetal force acting on the air in the jar. Newton's law says that for the same force, the product of mass and acceleration is the same. So if the mass is smaller, the acceleration must be more. Her explanation is, of course, technically correct, but for we lay scientists, it is simpler to comprehend that the centripetal force has more effect on the denser air than it does on the less dense flame.

9 AIR SPACE

Why, when a candle is standing in water and is covered by an upturned glass, does the water in the glass rise when the candle goes out?

This question was originally posed to *New Scientist* by three Australian schoolchildren, Emma, Rebecca, and

Andrew Fist of Norwood, Tasmania, who wanted to know why if they used four candles under the glass instead of one, the water rose even further up the upturned glass. They decided that what their science teacher told them—that the water rose to replace missing oxygen—couldn't be correct, and wanted to know why.

WHAT DO I NEED?
- four candles
- a large glass or jar capable of covering all four candles after they have been lit
- a bowl with a flat bottom
- water

WHAT DO I DO? Stand a single candle in a bowl. Firmly anchor it to the bottom so that it can't topple over. Fill the bowl with water to about an inch in depth. Light the candle. Place the upturned glass over the candle. Repeat the experiment using all four candles under the upturned glass jar.

WHAT WILL I SEE? When the candle goes out, the water level in the glass rises. Intriguingly, when you repeat the experiment using four candles, the water rises even higher than it did in the single-candle experiment.

WHAT'S GOING ON? It will come as a shock to many people who were shown this experiment as part of a school science lesson—and it might surprise a few teachers, too. Many of us were taught that the rising water level is caused by the oxygen under the glass being consumed by the burning candle and the water rising into the glass to replace the void created by the missing oxygen. It took three schoolchildren to question the orthodoxy associated with this classic experiment.

The consumption of oxygen by the burning flame may well contribute to the rising water level to a certain extent, because a given volume of oxygen will burn the candle wax's carbon into roughly the same volume of carbon dioxide and its hydrogen into twice the volume of water vapor. The former will partly dissolve into water, while the latter will almost completely condense into liquid water. This will lead to a net decrease in gaseous volume.

However, this is a minor consideration. Just as was the case in the "Sucking Eggs?" experiment in chapter 1, the important influence in what you see is the heat created by the burning candle or candles. And when they are covered with an upturned glass, several candles will increase the air temperature around themselves more than a single candle will.

As soon as the candle or candles go out, the surrounding air cools and contracts, allowing more water to enter the glass, and the ratio of contraction is directly proportional to the initial average temperature of the air volume under the glass. More candles lead to more heat, a higher temperature before they are extinguished, and therefore a higher water level upon cooling. The disparity in water depth between a single candle and four candles shows us that the absence of oxygen cannot be the key factor, because the same amount of oxygen is consumed in each case. The reason that water rises into the jar is due to a contraction of the cooling gases after the candles have been extinguished.

bowl

upturned glass

A.

Water rises.

B.

Water rises higher with four candles.

C.

Water can rise as high as in diagram B if a single thick candle is used.

Further proof that contraction of expanded gases is the reason for the rising water level comes from the fact that the water level rises only after the candles have been extinguished, at which point the remaining gases cool. If the water rose because the oxygen under the glass was being consumed, it would rise in a linear fashion throughout the period of burning.

Incidentally, a candle flame goes out after only a small percentage of the oxygen has been used up, because oxygen must be at its usual proportion (it forms 21 percent of air) to allow flames to burn freely. Any less oxygen and burning is quickly retarded. Therefore, the rise in water level is far too great to be explained simply by the water replacing only the missing oxygen.

One other factor must be taken into account: the effect is partly caused by the thickness of the candles under the jar. You can get some of a similar effect as the four-candle version of the experiment using a single candle of much greater thickness. The thicker the candle, the higher the water will rise. The water drawn into the glass must fit in the volume left between the candles and the glass and, obviously, the smaller this space, the higher the water will rise.

○ OIL LAMP

How can Alka-Seltzer tablets be used to create a 1970s-style "lava lamp"?

This is a trendy effect—or it is if you spent the late 1960s and '70s listening to psychedelic bands and wearing love beads. It works with any effervescent antacid tablet, not just Alka-Seltzer tablets.

WHAT DO I NEED?
- a clean, plastic two-liter soda bottle
- water
- food coloring
- vegetable oil
- Alka-Seltzer or other effervescent antacid tablets

WHAT DO I DO?
Dye the water with the food coloring—about ten drops ensures the water will be quite dark. You can choose the color of the water, but red or purple gives a pleasing effect. Fill a bottle three-quarters full with vegetable oil, and top it off with the colored water. Divide the antacid tablet into eight sections and drop one piece into the bottle.

WHAT WILL I SEE?
Oil and water don't mix, so at first the oil—which is less dense than the water—will float above the colored band of water at the bottom of the bottle. You'll also notice that the food coloring stays in the water and doesn't taint the oil. When you add the tablet, it will fall through the oil to the water below and begin fizzing as it reacts with the water. Then you'll see globules of colored water begin to rise up through the oil before reaching the surface and sinking again. When the reaction stops you can start it again by adding another piece of antacid tablet.

WHAT'S GOING ON?
Antacids are made from sodium bicarbonate and citric acid. When these are placed in

water they react vigorously. One of the key products of this reaction is carbon dioxide, plus water and various salts. So when the tablet reaches the water at the bottom of the bottle, it reacts with the water, forming bubbles of carbon dioxide. These bubbles become attached to the oxygen and hydrogen molecules that make up water and, because the bubbles are much lighter than water or oil, they rise to the surface, dragging colored liquid with them. Oil is viscous, so the bubbles' journey to the surface is relatively slow, which means they rise with a pleasingly gentle action. When they reach the surface, they burst, releasing the colored water they dragged along for the ride and allowing it to sink once more to the bottom of the bottle. It is this constant rising and falling of colored water that leads to the impressive lava lamp effect.

P.S. The original lava lamp was invented by Ronnie Rossi in the 1960s. It comprises a glass bottle sitting atop a lightbulb that heats the contents of the bottle. Inside the bottle are water and colored wax with a metal coil at the base. At room temperature, wax is denser than oil, but when it is heated it becomes less dense. When the lamp is switched on and the bulb heats the bottom of the bottle and its contents, the wax, which at room temperature sits at the bottom of the bottle, begin to heat up. As it becomes less dense, globs of wax begin to rise through the water. When a globule rises, however, it moves away from the heat source and cools, and its density begins to increase again. This means it falls back through the water toward the bottom of the bottle, where it starts to heat up again.

The metal coil also helps the wax to recoalesce so that the process can restart. The separated pieces heat up again and rise once more, creating a cycle as long as the heat-supplying light is switched on. A temperature of around 140°F needs to be maintained at the bottom of the bottle. If the system is too cold, the wax will simply sit at the bottom of the lamp; if it's too warm, it will float on the surface of the water.

○ DRINKER'S LEGS

What causes the thin, separated films that run down the inside of a glass containing wine, whiskey, or other spirits?

Known as "legs," these thin films are often considered the measure of a good spirit. You can be the judge of this if you try a variety of spirits and check out whether the good ones show the phenomenon best.

WHAT DO I NEED?
- a glass tumbler (or several)
- wine or whiskey (and other spirits if you wish to compare and contrast)
- water

WHAT DO I DO? Pour a decent measure of the alcohol and swill it around the glass. Do the same with a glass of water of the same volume. Take sips of both and watch the liquids drain down the sides back into the bottom of the glasses. You can try other spirits to see if they show a difference.

WHAT WILL I SEE? The thin film of wine or whiskey draining back into the glass after you have swirled it and sipped it will separate into rivulets, which will be roughly equally spaced. These are the legs. Water, however, drains back more rapidly. Because any film it creates on the glass falls back to the bottom quickly, legs do not develop, and the most you will see on the inside of the glass is a few beads of liquid.

WHAT'S GOING ON? This phenomenon was first correctly interpreted by James Thomson, the elder brother of Lord Kelvin (best known for his scale of absolute temperature measurement and his work on thermodynamics). In an 1855 paper titled "On Certain Curious Motions Observable at the Surfaces of Wine and Other Alcoholic Liquors," published in *Philosophical Magazine* (volume 10, page 330), Thomson explained that the effect is caused by capillary motion (capillary action), or surface tension, which was explained in the "Flower Power" experiment on page 69.

Alcoholic drinks are mixtures of ethanol and water. The ethanol—which has a lower boiling point than water—evaporates faster than the water when the drink has been poured. The surface tension of ethanol is lower than that of water, and the different rates of evaporation induce conflicting forces between the more watery regions of the drink and the more alcoholic regions. In effect, the watery and alcoholic regions are pulled apart, leaving legs of increasingly watery liquid as the ethanol evaporates and the watery remains start to trickle back down the glass under their own weight.

Thomson's paper won him little fame. This kind of

surface tension convection is now known as Marangoni convection after Carlo Marangoni, who worked on similar studies twenty years later.

Marangoni convection can cause other remarkable effects. For example, if the wine or whiskey is placed in a very flat dish and exposed to the air, the whole body of liquid becomes divided into hexagonal columns that circulate as the alcohol evaporates. Unfortunately, this is difficult to see in the home lab.

Normal convection in liquids, caused by density differences, usually swamps the effect of Marangoni convection, and it cannot get to work except in special circumstances, like on the side of your glass. However, in the weightless conditions of space, Marangoni convection rules supreme and is therefore a fertile topic for research by space scientists.

P.S. Legs in liquors and wines have been noted for millennia. The first account may even have appeared in the Bible: "Look not thou upon the wine when it is red, when it giveth his color in the cup, when it moveth itself aright" (Proverbs 23:31). It seems that even then the effect was considered an indicator of good-quality alcohol.

9 CREAM ON

Why does a thin layer of cream on Tia Maria create rapidly circulating and swirling patterns?

This is a perfect experiment for the after-dinner tabletop laboratory. Use a good light cream—milk is too runny, whipping cream too thick.

WHAT DO I NEED?

- a wide-bottomed glass
- light cream
- a bottle of Tia Maria coffee liqueur
- a spoon

WHAT DO I DO? Pour a quarter inch of Tia Maria into the glass and add a very thin layer of cream, poured gently over the back of the spoon so that it spreads over the surface of the Tia Maria.

WHAT WILL I SEE? The surface of the cream will break up into a pattern of cells as it reacts with the Tia Maria.

WHAT'S GOING ON? The swirling patterns are caused by convection—the bulk movement of fluids that is usually caused by temperature differences. However, in this case the convection is driven by a difference in concentration between the two liquids. This is called solutal convection.

The alcohol in the Tia Maria is the driving force. After the cream is added, the alcohol begins to diffuse through it. When it reaches the surface it alters the surface tension, reducing it. Areas unaffected maintain their higher surface tension and pull the low-surface-tension liquid areas toward them. As the surface liquid is pulled away, Tia Maria moves up to fill its place. This creates even less surface tension than the liquid pulled away. The process becomes self-sustaining and a convection cycle is created. This will last as long as a difference in concentration between the cream and the Tia Maria remains.

Surface tension convection is known as Marangoni convection. It plays a part in paint drying and causes patterns such as those seen in the "legs" running down the inside of a wineglass (as seen in the "Drinker's Legs" experiment on page 84). Similar effects are seen in rolling clouds or in oil frying in a pan. Tia Maria is something of an oddity, though, because in other systems the convection cells are circular or hexagonal.

P.S. As this phenomenon is one that *New Scientist* brought to the attention of the world, we feel particularly smug to include this experiment in the book. When the question was first asked in the magazine's "Last Word" column, nobody knew the answer, which led to a group of researchers studying the phenomenon and writing the definitive paper on the subject.

WANT TO READ MORE? The paper can be read in *Physica A*, 2002, volume 314, page 291.

⊙ OVER THE TOP

Sparkling wine or beer poured into a dry glass froths up spectacularly, but if the inside of the glass is wet, it doesn't. Why?

Quite frankly, the *New Scientist* research team was amazed at the number of home experiments you can do that involve alcoholic drinks. If you are interested only in experimenting, you don't have to drink the wine or beer, but . . .

WHAT DO I NEED?
- freshly opened sparkling wine or beer
- champagne flutes or beer glasses
- olive oil
- dust (look behind the wine rack)
- a tablespoon of sugar

WHAT DO I DO? Pour some of the wine or beer rapidly into a vertical glass so that it froths up to the rim. Let the bubbles subside and then add more.

WHAT WILL I SEE? The second batch of liquid will not have the same frothing effect as the first.

WHAT'S GOING ON? Beer, sparkling wine, and other fizzy drinks are supersaturated with gas. Although the laws of thermodynamics favor the gas bubbling out of the dissolved state, bubble formation is unlikely since bubbles must start small. Because the pressure of these tiny bubbles can reach about thirty atmospheres in a bubble only 0.1 micrometers in diameter, and the solubility of gases increases with increasing pressure (Henry's law), the gas is forced back into solution as quickly as it comes out.

However, bubbles can form around dust particles, surface irregularities, and scratches that serve as nucleation sites and repel water. (For more on nucleation sites, see the "Bouncing Rice" experiment in chapter 1.) Once the gas pocket around a nucleation site reaches a critical size, it bulges out and rounds up into a properly convex bubble, the radius of curvature of which is sufficiently large to prevent self-collapse.

Then there is a cascade effect. If the bubbles reach a certain critical number per unit of volume, this in itself constitutes a physical disturbance and results in the release of yet more bubbles.

Nucleation may be precipitated by a variety of imperfections. Minute salt crystals, such as calcium sulfate, may be present if the glass has been left to dry by evaporation after being washed in hard water. Or there may be tiny cotton fibers if the glass has been dried with a dish towel. Dust particles may have settled on the glass if it has been left standing upright on a shelf for any length of time. And tiny scratches will be present on the inside surface of all but brand-new glasses.

Once the inside of the glass is wet, any salt crystals will have dissolved and any cotton fibers will no longer function as centers of nucleation. Most of the dust particles and all of the scratches will, of course, still be there. However, these will have been coated with liquid, and the fresh carbonated liquid will reach them very slowly, by diffusion. Bubbles will still be produced, but at a rate that is too slow for the cascade effect to come into play. As a result, the drink will not froth over.

To demonstrate this, take a glass and thoroughly coat the inside with olive oil, which is a more efficient surface-covering agent than water. Then add the wine or beer (if you are feeling frugal, use a carbonated drink such as seltzer). The effervescence will be nil or minimal. Throw in the dust you scraped from behind the wine rack and see how it can affect the behavior of the liquid. Finally, add a few million centers of nucleation from a large spoonful of granulated sugar and prove to yourself how volcanic the effervescence can be.

P.S. Thanks to modern production techniques, today's glasses are of such good quality that some manufacturers build in deliberate imperfections, especially in beer glasses, in order to generate enough bubbles to maintain the head on the top of your tipple.

9 WHITE WATER DRINKING

Why do anisette-based drinks, such as ouzo, sambuca, and Pernod, turn white when water is added?

It seems strange that two completely colorless liquids combine to produce an opaque white one, immediately transforming something that was in solution and invisible into something insoluble and visible.

WHAT DO I NEED?
- cocktail glasses
- a bottle of ouzo, sambuca, or pastis
- a jug of water

WHAT DO I DO? Pour one of the anisette-based drinks or, if you wish to be a true scientist running controls alongside your main experiment (as described in "Fizz Fallacy" in chapter 1), a glass of all three. Add water.

WHAT WILL I SEE? The drinks will turn from clear to milky white (or, in the case of pastis, milky yellow).

WHAT'S GOING ON? Anisette-based drinks rely on aromatic compounds called terpenes for their flavor. These are soluble in alcohol, but not in water. The 40 percent or so alcohol by volume present in the drinks is enough to keep the terpenes dissolved, but when water is added they are forced out of solution as the alcohol content of the drink falls, producing a milky-looking suspension.

Absinthe, a similar drink derived from wormwood, gives a more impressive green suspension, because terpenes are responsible for a lot of the harsher plant scents and flavors found in the drink, including those of lemongrass and thyme.

P.S. Absinthe remains an almost mythical spirit reputed—but not proven—to have psychotropic effects. However, its toxicity means that it is still illegal in the United States. The Food and Drug Administration states that the presence of the chemical thujone in absinthe has harmful neurological effects, while absinthe producers claim thujone levels are insufficient to induce any response in humans. Interestingly, pastis was first produced by the distillers Pernod and Ricard after absinthe was banned in France in 1915. They removed the wormwood and added more star anise.

๑ YOU ONLY DRINK TWICE

Is there any taste difference in a vodka martini that is shaken, not stirred?

James Bond is famously fastidious in demanding his martinis be only ever served shaken. Apparently, there *is* a difference.

WHAT DO I NEED?
- a decent recipe for a vodka martini (The following one, in which olives and citrus fruit are added prior to shaking or stirring, is adapted from http://www.drinkoftheweek.com.)
- 3 shots (4.5 ounces) of vodka
- 8 drops of vermouth
- 2 olives
- twists of lime or lemon
- a jug
- 2 cocktail shakers
- martini glasses

WHAT DO I DO? Mix the vodka, vermouth, olives, and lime or lemon in a jug (gently, to avoid any undue shaking or stirring). Carefully divide the mixture into two cocktail shakers. Shake one, stir the other, strain, and then pour the martinis into two separate glasses.

WHAT WILL I SEE? Not a lot, except two exquisite cocktails, but you may—just may—be able to taste the difference. Ideally you'll need to recruit a volunteer for a blind tasting—don't tell them which is which.

WHAT'S GOING ON? Supposedly, when a martini is shaken not stirred, the vodka is "bruised." To seasoned martini drinkers this alters the taste.

However, there is great debate as to what is really happening. It is impossible to bruise alcoholic spirits in the way that you can bruise fruit (or indeed your arm), because they are liquids and have no vascular systems. The olive and citrus fruit (if you add them to the cocktail mix before shaking or stirring, as our recipe does) can be bruised by the shaking action, thus releasing flavorsome oils and juice. Other recipes eschew adding olives and fruit beforehand so that only the vodka and vermouth are shaken or stirred, thus ruling out botanical intervention. Try repeating this experiment, adding the olives and fruit after shaking or stirring, to see if you can taste the difference.

Amazingly, seasoned martini drinkers can detect a difference when the vodka and vermouth are shaken or stirred without the olives and fruit. This is because a martini is usually drunk within seconds of preparation, rather than minutes. The tiny bubbles caused by the shaking mean that a well-shaken martini is cloudy. This has an effect on the texture of the drink—it is less oily than the stirred version—and hence the taste is ever so slightly altered.

The bubbles from shaking can also partially oxidize the aldehydes in the vermouth in the way that the flavor of red wine is altered when we oxidize it—commonly known as letting the wine "breathe." This can alter the flavor of the martini—again, ever so slightly.

So now that you know, be honest—can you tell the difference between shaken and stirred?

P.S. Agent 007 has provided *New Scientist* with plenty of material over the years. One of the most surprising discoveries we made was that the kind of silencer that Bond uses when ruthlessly dispatching the bad guys is more myth than fact. It seems Hollywood has taken great liberties with silencers. Most real ones are much larger than the cigar tube–sized versions shown in the Bond films, and they are very difficult to fit and remove. If James Bond did manage to fit a silencer quickly enough, his gun would not produce the distinctive "phut" you hear in the films, but more of a firm crack, similar to the sound of a car door being slammed. For, despite what you see in the movies, it is almost impossible to silence a gun, because the gap between the cylinder and the barrel allows gas to escape. Sadly, because this is a book of experiments that you can try at home, we cannot recommend a trial of this particular 007 phenomenon. It is included merely on a whim to add to the ambience of your vodka martini experiment.

⟲ YO HO HO

Why does dark rum form a frothy scum on the top of the drink when mixed with cola, but white rum does not?

As you have by now discovered, there are quite a few opportunities to experiment with alcohol. Here's a chance to try one of the world's most popular drinks, the Cuba libre, while engaging in some serious scientific endeavor.

WHAT DO I NEED?
◆ a bottle of dark rum
◆ a bottle of white rum

- cola
- lots of tumblers (all good experiments are repeated to ensure consistency of results)
- good detergent

WHAT DO I DO? Pour a decent measure of white rum into one tumbler and a similar amount of dark rum into another. Top up the tumblers with cola (the less you use, the denser the resulting scum will be).

WHAT WILL I SEE? On the surface of the dark rum and cola you'll see a lacy network of darkish scum. Consume some of the drink and you'll see that the scum lines the glass in the way that a stale head from beer leaves a residue on the inside of your glass.

WHAT'S GOING ON? Dark rum is essentially white rum plus colorings and flavorings. One of the traditional flavorings used in dark rum is molasses—the residue left after the juice of sugar cane has been boiled, concentrated, and had as much sugar as possible extracted from it. Molasses is a good source of calcium and magnesium salts; it also contains tarry caramelized compounds and degraded organic substances such as waxes, fatty acids, and polysaccharides.

Cola (flavored solutions of phosphoric acid), when combined with the rum's solutions of alkaline earth ions such as calcium and magnesium, forms insoluble phosphates. In water you would probably see these as clouds, but in low concentrations such as in cola mixed with rum, fatty acids and waxes bind the tiny crystals and bubbles into a scum. This readily congeals into the material you see on the surface of the drink.

P.S. Remember to wash your Cuba libre glasses right after the experiment. The scum is notoriously difficult to remove if you leave it until the next day, and your mood will not be enhanced if you have consumed a great deal of your research the night before.

9 BEER ORDERS

If you are making a shandy or panache—beer plus lemon-lime soda or sparkling lemonade—and the beer is poured into the glass first, followed by the soft drink, the contents will fizz up and possibly overflow. But add the beer to the soft drink and this doesn't happen. Why?

This is a good ice-breaker at parties or in the bar. Kids can't drink the shandy, of course, but they can have fun pouring and then consuming the dregs of the soft drink afterward.

WHAT DO I NEED?
- tall beer glasses
- bottles of beer
- lemon-lime soda (or sparkling lemonade)
- paper towels for mopping up

WHAT DO I DO? Take one of the glasses and half-fill it with beer. Add the soft drink. Repeat the experiment with one more glass, but put the soft drink in first and then top it with beer.

WHAT WILL I SEE? The glass half-filled with beer will froth up as the soft drink is added, but the effect will be much less marked in the glass that is half-filled with the soft drink first.

WHAT'S GOING ON? If you take another two clean glasses and add only beer and lemon-lime soda or sparkling lemonade to each, you'll notice that the beer forms a longer-lasting head. This is because it contains substances known as surfactants (learn more about them in the "Bubble Nuts" experiment in chapter 1) as well as proteins and other long-chain molecules that help it to form and stabilize bubbles. Lemon-lime soda, on the other hand, forms bubbles that fizz and burst more quickly. The bubbles in sparkling lemonade burst even more quickly.

When you pour a soft drink into beer it rapidly sinks, hits the bottom, bounces back, and starts to bubble. Because these bubbles pass through the beer above them, they pick up the surfactants in the beer and form a froth on the surface. Beer poured into lemon-lime soda or sparkling lemonade sinks quickly in the same way, but the bubbles that push up from the bottom in this case pass through the soft drink, which contains fewer (or in the case of sparkling lemonade, no) surfactants, and pop quickly when they reach the surface. By the time enough beer has reached the top, the drink has settled down and little fizz is left.

4 IN THE STUDY
(OR THE OFFICE . . .)

⊙ FINGER TRAINING

My guitar teacher says that I should train my fingers to be more dexterous by laying my palms flat on a table and wiggling each finger in turn. I do, but why do I find it more difficult to lift my ring finger than any of the others?

This is a more useful exercise than the old school classroom trick of getting your friends to place their knuckles on the desk in the form of a fist and telling them you'll give them ten dollars if they can lift and fully extend all their fingers off the surface, one by one. (They can't; the ring finger stays stubbornly on the desk.) Both situations are affected by the same anatomical principles.

WHAT DO I NEED?
- your hand
- a friend's hand
- a flat, horizontal surface
- a ten-dollar bill that—all being well—you won't have to hand over

WHAT DO I DO? Place your palm flat on the surface and try to lift and wiggle each finger in turn. Then place the second knuckles of your fingers on the surface, palm downward in a fist, and try to lift each finger in turn, while keeping the knuckles of your other three fingers on the surface. Now, bet your friend ten dollars that he or she can't lift the ring finger from both of the positions either.

WHAT WILL I SEE? Your ring finger will be far less mobile than the others. And in the case of the knuckles-on-the-surface experiment, you'll find that, unlike your other fingers, you can't lift it at all.

WHAT'S GOING ON? The anatomy of the human hand is very complicated. An "extensor" muscle extends or straightens a digit when it contracts, while a "flexor" bends it. Extension of a single digit requires the simultaneous contraction of the extensor muscle for the appropriate digit, the relaxation of the flexors for that digit, and the contraction of the flexors for all the other digits to make them stay where they are.

This happens fairly easily in the thumb, index finger, and little finger, because they each have their own designated extensor muscles. However, the middle and ring fingers use a common extensor muscle. When attempting to extend the ring finger, the middle finger flexor contracts as it is supposed to, effectively pinning down the middle finger, but the two digits' common extensor then pulls against it to no avail. Try lifting the middle and ring fingers as a unit and see how easy this becomes when the middle finger no longer acts as a tether.

Because the hand evolved primarily for grasping, playing a musical instrument can push its design specifications to the limit. Researchers at Cardiff University in Wales, studying the biology of musical performance, have shown that there is a high degree of variability in the muscles and tendons of the hand, and it seems that the success or otherwise of musicians can often be the result of differences in this physiology. Because of this, not everyone can make the same movements. For example, in about 20 percent of people, curling up the thumb causes the index finger to bend inward because of an anomalous link between flexor tendons. This renders some piano and guitar fingerings impossible for those affected.

On the other hand, evidence from touch typists shows that their highly trained and flexible fingers can often move separately with no difficulty, while younger people who spend a lot of their time texting have astoundingly flexible thumbs.

The link between the middle and ring fingers is almost universal, but a very few people have aberrant connections, so if you can't afford to lose ten bucks, beware of the rare exceptions and those other practitioners of finger flexibility, the touch typists.

P.S. *New Scientist* reader Conor Nugent has a different take on the immobility of the ring finger, telling us, "The ring finger has the least dexterity because it is used the least. The index finger is used for pointing, the little finger is extended while drinking tea, and the middle finger is used when driving. The ring finger, however, is used only once, shortly before marriage."

○ SEE BETTER

Why does looking through a pinhole allow you to see objects more clearly?

I noticed this effect in my youth as I became progressively nearsighted and was trying to peer at the blackboard in classes and lectures. (Sadly, I didn't find a similar cure for early onset baldness.) What I discovered was that if you peer through a small gap, such as that formed between curled-up fingers, objects that were blurred become clearer. The effect is particularly marked if you are nearsighted and remove your glasses or contact lenses before you try this out.

WHAT DO I NEED?
- a piece of card stock or thick paper
- a pin
- paper with words written on it, much like an optometrist's eye-testing chart
- the human eye (preferably a myopic one)
- a comb

WHAT DO I DO? Punch a tiny hole in the card stock with the pin. Fasten your homemade eye-testing card to a wall at a distance where you find it difficult to read. Peer through the pinhole in the card with one eye while keeping the other eye closed.

WHAT WILL I SEE? The writing on the card stock will become clearer when you look at it through the pinhole. You'll notice the same when you look at other objects around you and in the distance. The effect is particularly noticeable if you are nearsighted, hence our earlier suggestion that if you normally wear glasses or contact lenses, remove them before attempting this. If you have excellent vision, the effect will not be as obvious, but there has to be some compensation for nearsightedness, doesn't there?

WHAT'S GOING ON? Under normal circumstances, light rays entering the eye's lens are not focused in one place. For you to see a clear image of a given point, the lens has to concentrate all the rays into a single point on the retina. When the eye is not perfectly shaped, as in nearsighted people or those with an astigmatism, the outermost rays entering the eye are not bent or refracted by the lens enough for them to be focused on the correct point at the back of the eye. The innermost rays entering the eye do not need to be bent as much to hit the middle of the retina, and these travel a relatively straight route to form a clear image, even in people who are nearsighted. However, the outermost rays confuse and blur the image. When you look through a pinhole, the bundle of rays entering your eye is greatly reduced, allowing only the inner rays that pass through the central portion of your lens straight to your retina to enter and excluding the peripheral rays, which cause the blurring. This means you see images clearly. The drawback to the improvement in vision is that the total amount of light entering your eye is greatly reduced by the pinhole, so images seem darker and, in low light, the reduction in brightness offsets the increased focusing gains.

Try using the gaps between the teeth of a comb to

reproduce this effect. Native Alaskans have historically worn glasses with narrow slits in them, reproducing the effect of looking through a comb. More important, because snow and ice reflect a lot of light, which can produce excessive glare, looking through slitted glasses helps to reduce the amount of light entering their eyes, aiding vision and preventing snow blindness.

P.S. Pinhole glasses, similar to slitted glasses, use this effect to correct vision. They have a number of benefits over normal prescription lens glasses, the most important being that as your eye changes with age, the pinholes do not need to be altered. One pair of glasses will last a lifetime. The obvious drawback, however, is that they greatly reduce peripheral vision.

⑨ COLOR CHANGE

If you stare at a color for long enough, it seems to impinge on your vision. How?

Many of us have experienced this after looking at a bright light or having our photograph taken with a camera flash in use. You spend the next five minutes or so blinking, because you can still see the image of the light, even when you close your eyes. Why do you still see something that's no longer there?

WHAT DO I NEED?
◆ four large sheets of white paper (8½ × 11 inch sheets will do)

- pieces of bright red, blue, and green card stock or thick paper
- scissors
- glue
- a black marker pen
- a well-lit room

If you want to try the clever bit described at the end, you'll need:
- more sheets of paper
- green, yellow, and black crayons

WHAT DO I DO? Draw a fishbowl on one of the large sheets of white paper with the black pen. Cut out three identical fish shapes (large ones, but smaller than the bowl you have drawn) from each of the three colored pieces of card stock and use the glue to stick these to the three remaining sheets of white paper, one on each. Hang all four sheets of paper on the wall of a brightly lit room.

WHAT WILL I SEE? Stare at the red fish for about thirty seconds, then stare at the fishbowl. You should see a greenish blue fish in the bowl. Look away, let your eyes return to normal, and repeat what you have just done with the green fish. This time you'll see a reddish blue fish in the bowl. Repeat again with the blue fish and you'll see a yellow fish in the bowl.

WHAT'S GOING ON? The retina of the human eye is covered with light-sensitive cells called rods and cones.

The rods, of which there are more than a hundred million, pick up on shading and levels of light and dark, while the cones, of which there are only around seven million, register color. There are three types of cone, each sensitive to a different range of color—red, blue, or green. Though there are far fewer cones than rods, the cones are concentrated especially in the central area of the retina known as the fovea. Cones are more active at higher light levels, while rods are responsible for vision in dusk and in the dark—hence the requirement for a brightly lit room to ensure this experiment works at its best.

The images you see in the bowl after staring at the fish are afterimages, but why do these appear in a different color from the one you were looking at? When you stare at the red fish, the image falls onto one region of your retina, and the cells responding to red become desensitized. White paper reflects all the colors of the spectrum rather than absorbing any one of them, unlike the red, green, and blue cards from which the fish are made; these instead absorb colors, hence their appearance. So, when you turn to look at the fishbowl on the white paper after staring at the red fish, the red-sensitive cells don't respond as strongly as normal to the red part of the light that is being reflected back from the white surface, yet the cells that are sensitive to blue and green light do. That's why you see a greenish blue fish.

Obviously, if you stare at the green or blue fish before looking at the white paper, the green- or blue-sensitive cells become fatigued and the fish in the bowl changes color accordingly.

But what causes the afterimage in the first place? The old explanation was that the cone cells become tired when they have to constantly "fire." In fact, the afterimage is due to the nerve cells in the eye that process the signals from cone cells. When they keep receiving a red signal from the cones, they start to adapt by turning down the "volume" of the signal they send to the brain.

When we look away from the image, the red signal from this part of the eye stays muted for a while. However, the blue and green signals are sent at the normal volume. Your brain doesn't appreciate that the red signal has been turned down and interprets this as a normal signal from your eyes. This means that the fish-shaped patch from which the red signal is reduced is perceived as a genuine object in your field of vision, leading to an afterimage that persists even when the object has been removed.

Normally, the eye deals with this problem by shifting back and forth slightly when you think that you're staring straight, so that the same cones are not constantly exposed to the same image and there is no desensitization. However, if the image is large enough, like the fish you have been staring at, these small eye movements, known as saccades, are not enough to change the color being detected by one part of the eye, and, as we have seen, this part will stop responding as strongly as normal.

P.S. A clever way to test this phenomenon is to draw an American flag on a piece of paper, replacing all the red with green, all the blue with yellow, and all the white with black. Stare at the flag for thirty seconds and then stare at a blank white background as before. The flag will appear as an afterimage in the correct colors.

9 SEALED IN LIGHT

Why do self-sealing envelopes appear to spark when opened in a darkened room?

It's important to make sure you use a self-sealing, gummed envelope to see this. The good news is that it will work repeatedly on the same envelope—just reseal it after opening and pull open the flap again.

WHAT DO I NEED?
◆ self-sealing envelopes
◆ a darkened room

Sugar cubes and a pair of pliers can be used to create a related effect.

WHAT DO I DO? Draw the curtains and switch off the lights (preferably at night). Let your eyes adjust to the darkness. Pull open some self-sealed envelopes, making sure the two gummed and sticky pieces of paper keeping the envelope closed are pulled apart as rapidly as possible without tearing the rest of the envelope.

WHAT WILL I SEE? A fluorescent glow, or "sparking," will appear along the line of the gummed surfaces as they are torn apart.

WHAT'S GOING ON? The colored glow is a form of chemiluminescence—the emission of light without heat as a result of a chemical reaction. Separating the gummed surfaces requires energy that breaks the attractive forces between the molecules of gum.

The act of pulling apart the surfaces supplies excess energy to the gum molecules, which lifts them into an excited state. As they decay back to their normal state, energy is released in the form of visible light. The difference in energy between the excited and ground states defines the wavelength and hence the color of the light produced; in this case, purple.

Chemiluminescent glow can also be seen when stripping lengths of electrical insulating tape, and although people have speculated that it might be responsible for some gas-induced mining explosions, there is no evidence to suggest there is enough energy in the sparks to ignite the methane present in mines.

This phenomenon is different from fluorescence, where light (often invisible ultraviolet light) is absorbed and then reemitted at a longer wavelength in the visible spectrum. Fluorescence gives rise to loud, brash, glowing colors such as the strange blue glow you observe while drinking tonic water near one of the ultraviolet lamps sometimes found in nightclubs.

P.S. You can see a similar effect if you take a pair of pliers and crush sugar cubes in a darkened room. When sugar crystals are crushed, tiny electrical fields are created, separating positive and negative charges that then create sparks when trying to reunite. Here the phenomenon is known as triboluminescence—light generated by the breaking of bonds in a crystal.

9 ELASTIC FANTASTIC

Why do rubber bands heat up when you stretch them?

The next time someone flicks a rubber band at you, don't shoot it back at them—at least, not right away. First, take a moment to try a simple experiment that reveals a strange behavior of this everyday object.

WHAT DO I NEED?
◆ a thick rubber band
◆ a (relatively) hairless upper lip (there's no need to wax)

If you want to study the effect further, you'll need:
◆ a skipping rope
◆ a friend

WHAT DO I DO? Stretch the rubber band between your hands and extend it as far as it will go. Now hold it against your hairless upper lip and let it contract quickly. Wait a moment, then, keeping it against your lip, quickly extend it again.

WHAT WILL I FEEL? You'll notice that as you let the rubber band contract, it feels cool. As you extend it, the rubber feels hot. Quickly release the band and it will feel cool once more.

WHAT'S GOING ON? Rubber is made up of a tangled network of extremely long, flexible polymer chains—a bit like a crumpled fishnet—that are constantly vibrating and rotating. Polymers are substances with molecular structures built up mainly from a large number of similar units bonded together.

As you stretch a rubber band, these polymer chains align and are pulled closer together, reducing the range of motions they can undergo. This reduces their entropy (the degree of disorder or randomness they possess). According to the laws of thermodynamics, equilibrium must be maintained. So entropy has to increase in some other way to counteract this reduction, and the atoms on the chains compensate by vibrating more vigorously. As a result, the rubber heats up.

When the rubber is allowed to contract, the process reverses—the chains become free to rotate and vibrate, and their entropy or randomness increases. To compensate, the entropy of the atoms in the chain decreases, the atoms vibrate less, and their temperature falls.

Now you can shoot the rubber band back.

P.S. Rubber's unusual structure means it behaves quite differently from other materials. If a length of stretched rubber is heated, for example, its polymer chains will begin to vibrate more vigorously. This tends to pull the ends of the chains toward each other—in other words, the rubber will contract. This is the opposite of the usual thermoelastic effect exhibited by metals, for example, which usually expand when heated.

The effect is easy to demonstrate. Hold one end of a skipping rope and ask a friend to hold the other end. Begin to turn it as you would if somebody were skipping. You'll notice that the faster the rope turns, the harder it is to hold the two ends apart. This is how rubber behaves, pulling in the ends of its polymer chains as it vibrates more vigorously under heating.

⊙ RUBBER HORROR

Why, when left for long enough, do rubber bands seem to melt and turn into a sticky mess?

After trying out the "Elastic Fantastic" experiment on page 110, put the rubber bands aside for a few months, preferably in a sunny place, and see how they change.

WHAT DO I NEED?
- rubber bands
- a sunny indoor spot such as a window ledge or a table near a window
- greaseproof paper to protect your chosen surface

WHAT DO I DO? Leave the rubber bands on the grease-proof paper for a few weeks or months in full sun.

WHAT WILL I SEE? Over the weeks and months, the bands seem to melt, at first becoming a sticky mess and then, when left even longer, often becoming brittle and crumbly.

WHAT'S GOING ON? Natural rubber is made up of polyisoprene chains that slip past one another when the material is stretched. When raw, the substance is too sticky and soft to be of much use, so it is toughened with chemicals such as sulfur. These create cross-links between the chains, making the rubber stiffer and less sticky. This process is called vulcanization.

Over time, ultraviolet light and oxygen in the air react with the rubber, creating reactive radicals that snip the polyisoprene chains into shorter segments. This returns the rubber to something like its original soft and sticky state—what you see when your bands have turned gooey. Later, the radicals form new, short cross-links between the chains. This hardens the rubber and eventually turns it brittle, leading to the crumbly end product. Any vulcanization agents left in the rubber contribute to this process.

Whether a rubber band is going to become sticky or brittle depends on the relative rates of these processes, and these rates in turn depend on the rubber's quality— what additives, fillers, and dyes it contains and whether these absorb light or help transfer radicals. The presence of strong oxidizers, such as ozone, in the air creates even more radicals. So the eventual fate of your rubber band depends in part on whether you have a desk next to an ozone-emitting machine, such as a photocopier.

Heat and light also speed up the reactions. The polymer chemistry of rubber is fairly complex, and so it is difficult to determine precisely how much heat and light is required to change the process. Obviously, the chemical reactions take longer if the rubber bands are left in a fridge than on a warm, sunny window ledge. A rule of thumb is that the reaction rate roughly doubles with a rise of 18°F in temperature, but this is more complicated when you take oxygen and light into consideration.

The final factors that influence the change are ultraviolet light intensity and whether the band is stretched or not—stretching brings the polyisoprene chains closer together, allowing radicals to jump from one chain to another more easily and thus create new bonds between the chains.

P.S. The biggest consumer of vulcanized products is the automobile industry, which purchases almost 80 percent of all new rubber, mainly to produce vehicle tires. The world produces more than a billion tires annually, with an estimated six billion in use in North America and three billion in Europe at any one time.

⊙ RIP OFF

Why does newspaper have a preferential direction for tearing?

While newspaper looks homogeneous, when you try to tear it from top to bottom or from side to side there is a noticeable difference. And newspaper isn't the only everyday object that has a grain.

WHAT DO I NEED?

◆ a newspaper

If you are happy with the results of your experiment, you can also buy some uncooked hamburgers (a combination of mass-produced and handmade ones helps to display the difference)—and hang on to the plastic bag you carry them home in.

WHAT DO I DO? Rip one sheet of newspaper along its length, then rip another sheet along its width.

WHAT WILL I SEE? One of the pieces will, with little difficulty, tear parallel to the edge of the paper. The other will rip haphazardly.

WHAT'S GOING ON? Most paper is made on machines operating at high speed. The sheets are made by draining a dilute suspension of fibers and mineral fillers on a table of continuous and fast-moving synthetic, sievelike wire. The sheets are then consolidated by pressing and finally dried on heated cylinders before being put onto reels.

As the papermaking suspension is discharged from a vessel known as the flow box or head box onto the rapidly moving wire, most fibers—which are cylindrical in shape—are aligned in the direction of the wire's move-ment. This is called the "machine direction." The orien-tation of fiber in the paper structure allows the sheet to tear more easily in the machine direction than in the "cross direction," where the fibers are presented sideways.

The strength of all paper made in this way is in-fluenced to varying degrees by the directional properties of the sheet. The addition of fillers and mechanical or surface treatment during manufacturing tends to modify or reduce the strength. As it contains mostly fiber and is lighter in weight, newsprint tears more easily in the machine direction than computer or photocopier paper.

P.S. You can look for a machine direction in any number of household items and foodstuffs—many possess similar qualities. Hamburgers cooking at the *New Scientist* summer barbecue were seen to change shape from round to oval due to similar manufacturing circumstances.

Commercial hamburger patties are made on a large scale—their ingredients are minced, extruded, rolled out, flattened, and stretched on a moving belt to give them the right thickness and texture before the patty discs are punched out by a machine. It is this extrusion and rolling that makes the hamburgers anisotropic (meaning their physical properties have different values when measured in different directions), because their protein fibers tend to get stretched along the direction of the belt of minced meat.

When cooking starts, the heat makes the protein expel water and shrink. The stretched fibers mainly shrink lengthwise, squeezing the circular patties into elliptical shapes—until you eat them. By contrast, handmade hamburger patties are isotropic (meaning their physical properties have the same values when measured in different directions) and cooking them will not alter their shape in any particular direction.

Many other rolled or extruded products display anisotropy. The plastic used to make grocery bags is extruded and stretched and so tends to have a preferred tearing direction. The bags are designed so that when loaded vertically and carried by their handles they are at their strongest.

⊃ IMMOBILE MOBILES

Does talking on a cell phone really slow your reactions?

It's a hot topic right now because in many states, driving while talking on a handheld cell phone has been made illegal. Skeptics seem to think that such legislation is an overreaction, and often you'll see people flouting the law. Here's a way to judge for yourself just who is right.

WHAT DO I NEED?
- one or more experimental subjects (preferably people who think their driving is unaffected by talking on a cell phone)
- a yardstick (any rodlike instrument that is divided into one-inch increments will do)
- a table
- a calculator

WHAT DO I DO? Ask the subject to place his or her forearm on the table with their open hand overhanging the edge and with the index finger and thumb at the top as if they were about to shake hands with someone. Hold the yardstick vertically just above the hand, with the bottom of the yardstick level with the index finger. Ask your subject to catch the yardstick as quickly as possible after you drop it. Note how many inches have slipped through before it is caught. Give three tries and take the average. Now ask your subject to call a friend on the phone and enjoy a good chat. When the conversation is in full swing, drop the yardstick. Now record where it is caught.

WHAT WILL I SEE? You will see a crude measure of how a telephone conversation can affect reaction time. And you'll find that talking to a friend can seriously harm your ability to respond to a change in circumstances.

WHAT'S GOING ON? Most researchers who study multi-tasking agree that holding a conversation interferes with a person's ability to react to events, especially those occurring on the road. When two stimuli are presented simultaneously, the reaction to one of the tasks will be delayed. And of great significance to all road users—drivers and pedestrians alike—is that you can't always choose which. So when your friend asks you something that needs a reply and a child steps into the road in front of you, there is a moment's hesitation as your brain attempts to make sense of the situation and select which is the more important issue to act on. This is known as the brain's "response selection bottleneck" and was first noted by researchers in 1952. The decision process may seem short, but at the speed a car travels it can easily make the difference between life and death.

P.S. To get the person's reaction time in seconds, divide the number of inches recorded by the falling measure by 193.2 and take the square root. (If you decide to use the scientists' standard and record the lengths in centimeters, you would divide the number of centimeters by 490 and take the square root.) Results taken from a number of volunteers in the *New Scientist* office provided us with average figures of 6.7 inches when volunteers were concentrating and 16.14 inches when they were being distracted by somebody talking to them on the telephone. These distances correspond to reaction times of 0.186 seconds and 0.289 seconds respectively, meaning people on the phone took 0.103 seconds longer to react—an increase of 55 percent.

This gives an idea of how people's reactions are delayed when they are talking on the phone in a quiet office, but on the road there is even more to distract a driver, including having a conversation with a fellow passenger or listening to the radio, both of which can increase the delay. Because a car traveling at sixty miles per hour travels eighty-eight feet in one second, the delay noted in our example could be crucial. The distance the car travels in this time is simply its speed multiplied by the time. In this case, this would be an extra nine feet—nearly a car length, and possibly the difference between life and death.

○ SPARK OF GENIUS

How does an electrophorus work?

An electro-what? This may sound like something Dr. Zachary Smith would have used to foil the Robinsons in *Lost in Space*, but it's actually a simple device that demonstrates, quite spectacularly, the effect of stored electrical energy. You can make your own version at home.

WHAT DO I NEED?
* a tack
* a flat-bottomed, aluminum foil dish (the kind used to hold prepacked pies or quiches)
* a pencil with a rubber eraser on the end
* superglue
* a polystyrene plate

- wool (you can use a sweater, but a carpet has been known to work)

WHAT DO I DO? Press the tack through the bottom of the foil dish so its point is sticking up into the middle of the inside of the dish. Press the eraser end of the pencil down onto the point of the pin so that the pencil juts—point facing upward—from the base of the dish. A little superglue will help to hold it firmly in place. You should be able to lift the whole assembly using just the pencil. (If you can't get this to work, glue a polystyrene cup to the foil dish and use that to lift it.)

Now rub the polystyrene plate with the wool for at least three minutes. Place the plate on a table or work surface and, holding only the pencil handle, position the aluminum dish on top of the polystyrene plate. Touch the dish once with your finger while it sits on the poly-styrene plate. Now lift the dish off the plate, again using only the pencil, and touch the dish again.

pencil

rubber eraser

foil tray

polystyrene plate   tack

WHAT WILL I SEE? The first time you touch the dish (that is, before removing it from the polystyrene plate), a spark may fly from the dish to your finger, and you'll certainly experience a small electrical shock. When you remove the dish from the plate and touch it again with your finger, you'll see another spark and really feel the electricity jump from the dish to your skin. If you want to see the sparks produce spectacular results, try the experiment in a darkened room.

WHAT'S GOING ON? Rubbing the polystyrene plate with wool creates a negative static charge on the surface of the plate by attracting electrons (which are negatively charged). This is called triboelectrification. By placing the aluminum foil dish on the plate using the eraser of the pencil, you are ensuring that no charge is lost through the dish because it is not grounded through you—you are holding the wooden part of the pencil and the pencil is stuck to the dish by insulating rubber. When the aluminum foil dish makes contact with the polystyrene plate, its electrons are repelled from its base because the negative charge on the electrons in the polystyrene, created by rubbing it with the wool, repel the negatively charged electrons in the foil dish. This is because these like charges repel each other in a similar way that two negative magnets repel each other.

This makes the bottom of the aluminum foil dish positive and the top of the dish negative. So, by touching the foil dish with your finger while it is still resting on the polystyrene plate, you allow these negative electrons that have been forced to the top of the dish to flow into you, creating a spark and an electrical shock that you feel in your finger. Because you have removed the negatively charged electrons from the foil dish, the dish is now positively charged. If you remove it from the polystyrene plate using your insulated pencil handle and touch its positively charged surfaces again with a finger of your other hand while it is suspended, you'll create another spark and shock as the foil dish is made neutral again by your grounded body.

P.S. You have made a device called an electrophorus, perfected in 1775 by Alessandro Volta, who was responsible for the development of the electric battery and after whom the electrical volt is named. The electrophorus was adopted by physicists around the world because it provided a source of charge and voltage for research into electrostatics.

○ SEEING DOUBLE

If you place one mirror facing another and stand between them, can you see infinity?

This question is perhaps more philosophical than scientific, but there's no doubt that when you look into one or another of the mirrors your reflection seems to stretch away into the far distance. So does it actually have an end?

WHAT DO I NEED?
+ two mirrors, the bigger the better

WHAT DO I DO? Place the mirrors facing each other as parallel as possible—standing them along two facing walls of a room will help you align them better. Stand between them and look at your reflection in either mirror.

WHAT WILL I SEE? Your reflection will seem to stretch into the distance in both mirrors, curving away to become smaller and smaller until you can no longer determine where it ends.

WHAT'S GOING ON? Sadly, you can't see infinity. No mirror reflects 100 percent of the light falling on it. If it's a very good mirror and can reflect 99 percent of the light, after about 70 reflections only 50 percent of the light will remain, after 140 reflections only 25 percent of the light is left, and so on until there is not enough light left to reflect between the two mirrors. Additionally, most mirrors reflect some colors of light much better than others, and some colors are absorbed better by the glass, so the multiple reflections you see not only get darker but also become more color-distorted as they recede.

Then there is the problem of mirror alignment. Even with perfect reflection of all colors, you could never see infinite reflections for geometric reasons. The faces of the two mirrors would need to be perfectly parallel. This is practically impossible to achieve. There will always be a slight disparity in their positions relative to each other. This is why the images appear to curve away, until eventually the reflection is lost "around the bend."

Even if the mirrors are perfectly reflective, perfectly parallel, and really huge, your eyes are in the middle of your head, not at its edge. Therefore, at some point, the receding and hence apparently more distant mirror images would become smaller than, and hidden behind, the first reflected image of your head. Even with a tiny camera and a giant pair of mirrors, the reducing reflection size would eventually be smaller than the first reflected image of the camera apparatus.

However, for those who prefer a theoretical mathematical explanation of whether infinity awaits us at the end of a reflection, there is hope. Assuming perfectly parallel mirrors, perfect reflection, and a transparent viewer who was prepared to stand around forever, infinity is just about achievable.

Take a deep breath . . . here's how it works. Light travels at a finite speed (c), which you measured in the "Hot Chocolate" experiment in chapter 1. (If you haven't done that experiment, light travels at roughly 3×10^8 meters per second.) If your two mirrors were L meters apart and you stood between them for t seconds, you would be able to see c multiplied by t divided by L reflections. For example, if your mirrors were two meters apart and you stood between them for one minute, you should be able to see about nine billion reflections. If you could stand around forever, those reflections would presumably reach infinity.

P.S. If you are nearsighted, you can take advantage of that to set up a study of another interesting property of mirrors. You may have to shift them around a bit to get one in a position where it is reflecting a distant object. Then take off your glasses or remove your contact lenses and you'll notice that although you are standing close to the mirror, the distant object that is being reflected is as blurred as if you were looking straight at it. This may seem odd because you are very close to the mirror.

The reason the object is still as blurred as if you were looking directly at it is because the light still has to travel the same distance (plus a little bit more) to your eye. It travels from the object to the mirror and then onto your eye, so you are looking at the sum of the two distances, not at the mirror itself.

A simple experiment to prove this is to get a small sticker and place it at eye level on the mirror. You'll notice that looking at your reflection, looking at the reflection of an object behind you, and looking at the sticker on the mirror all require you to adjust the focus of your eyes accordingly. It appears the object behind you is actually behind the mirror and exactly that same distance away from it. This proves that your eyes do not focus on the surface of the mirror but on the objects reflected in the mirror and the distances these objects are from the mirror.

Anybody who is acquainted with photography will understand these principles. When using a camera to focus on an object in a mirror, photographers have to calculate the effective distance by measuring the distance from the camera to the mirror and adding it to the distance from the mirror to the object. Some autofocus cameras have particular trouble with this process. They read only the distance from the camera to the mirror and need to be manually adjusted to compensate for this.

There are also lessons to be learned if you are in the habit of checking your appearance in the mirror. Remember that you are looking at yourself from twice the distance to the mirror. Which means that if you are checking for blemishes or stray pieces of hair, you need to halve the distance between yourself and the mirror to see yourself as others do!

5 IN THE BATHROOM

9 THE STUFF OF LIFE

I've been told you can extract your own DNA at home. Is this true?

Of course it is. And, as seems to be the case in so many *New Scientist* experiments, this involves a stiff drink. However, there's a perfectly good reason for breaking out the vodka—you can actually find out what makes you unique.

WHAT DO I NEED?
* some dishwashing liquid
* three teaspoons of water
* a small, clean glass
* a teaspoon of salt dissolved in a glass of water
* an ice-cold alcoholic spirit of greater than 50 percent alcohol by volume (cask-strength malt Scotch whisky, a quality gin, strong vodka, or rubbing alcohol should do the trick)
* an eyedropper

WHAT DO I DO? Put a teaspoon of dishwashing liquid diluted with three teaspoons of water into the clean

glass. Swish the salty water around your mouth vigorously for thirty seconds or so, then spit it into the
diluted dish soap. Stir this firmly for a few minutes,
then very gently pour a couple of teaspoons of ice-cold
strong alcohol down the side of the glass. Use the eyedropper if you don't have a steady hand; tilting the glass
also helps. This stage requires great concentration and is
very important as you must have a clearly demarcated
water/alcohol boundary. If you are careful, the spirit
will form a separate layer on top of the salt/spit mix.

WHAT WILL I SEE? Wait a few minutes and you'll see
spindly, white threadlike clumps starting to form in the
alcohol. This is your DNA.

WHAT'S GOING ON? Swishing salty water around your
mouth removes cells from the inner surface of your
cheeks in the way you've seen in TV dramas when the
police take a swab from a suspect's mouth for DNA
analysis. The detergent in the salt/spit mix breaks down
the cells' membranes, releasing the DNA in the cell
nuclei. Because DNA is soluble in water but not in
alcohol, it precipitates out in the white clumps you see
floating on the surface. If you have a microscope you
can investigate further, but if you just want to sit back
and admire it, that stuff floating in the glass is what
makes you who you are.

P.S. Because the spirit has to be chilled, vodka lovers will
be happy enough to contribute to this experiment, but
Scotch aficionados may turn their noses up at putting
their favorite spirit in the fridge. The Scotch drinkers

will either have to swallow their pride, buy a bottle of vodka or gin, or simply slum it with rubbing alcohol.

Before you start, make sure you have a clean mouth—if you've been scoffing a meat sandwich, it might not be your own DNA that ends up in the glass.

9 TESTING THE WATERS

Can we trick our sense of touch?

This popular experiment shows us that much of what we perceive through our senses is relative. We adapt to our changing circumstances.

WHAT DO I NEED?
* three bowls large enough to put your hands in, one with hand-hot water, one with tepid water, and one with cold water chilled in the fridge
* your hands

WHAT DO I DO?
Simultaneously place one hand in the cold water and the other in the hot water. Hold them there for ninety seconds. Then place both simultaneously into the bowl of tepid water.

WHAT WILL I FEEL?
When your hands are respectively in the cold-water bowl and the hot-water bowl, one feels cold, the other hot. Hardly surprising. Yet, when you place them in the tepid water they still feel different, even though they are both now in water of the same temperature. The hand that was originally in the hot water now feels cold, but the hand that was orig-

inally in the cold water feels warm. After both have been held in the tepid water for a while the sensations ease off, and eventually they feel the same.

WHAT'S GOING ON? Human senses are relative, measuring only the differences between things that we perceive. So, your cold hand registers the tepid water as warmer, and conversely your hot hand perceives the tepid water as colder. Your senses do not make absolute judgments on what is around them, only the relative differences of their environments.

While this may lead to a few disconcerting moments, as your hands have just experienced, it helps us focus on what is important and what is changing in the environment around us, and allows us to ignore the things that are neither. It's a process called adaptation and is not just a product of our sense of touch. It affects all our senses and helps to prevent them from suffering from overload. Without this process we would find the world a debilitating place.

Adaptation is the reason we get used to smells and stop noticing them, and why our eyes become used to brightness or darkness after we have been exposed to them for a little while. People who work in foul-smelling environments, such as in trash-collecting facilities or fish-processing plants, are aware of the odor when they first arrive at work, but after a few minutes the impact is reduced as their sense of smell adapts to their surroundings. Similarly, people who work in noisy environments are able to eventually filter out the constant background noise and hear what their colleagues are saying.

P.S. Adaptation may be the reason why, as we age, we believe that in our youth the sky was bluer, summer days were warmer, and food tasted better. This was studied in the 1950s when researchers sat volunteers in front of a screen and projected colored lights onto it for five seconds, and then again after a lapse of five more seconds. The volunteers were asked to adjust the controls on a colored light generator to try to reproduce the color they had just seen. The colors they chose were always brighter or deeper than those they had originally been shown. This is because adaptation was acting on their visual sense. They were given longer than the five seconds that they had viewed the original color to set up their matching color, so the longer they looked at the screen as they attempted to set up their match, the more the color appeared to fade as their sensitivity to it declined. This meant that they continually turned up the brightness or depth of the color as they worked at matching it to the original. The researchers believed that, because adaptation probably acts in the long term as well as the short term, the summer skies of our childhood seem brighter, the days toastier, and the roast chicken more succulent. Other researchers dismiss this idea, saying that the aging of our faculties is to blame. However, either argument might explain why Brussels sprouts, particularly distasteful in childhood, never taste as bad in adulthood. (Though some of my colleagues in the *New Scientist* office dispute this heretical suggestion, insisting that they are as disgusting as they ever were.)

○ AROMATIC PEE

Why does urine smell peculiar after you've eaten asparagus?

This really is quite a surprising effect, the onset of which can occur within minutes of eating asparagus. Once you recognize the smell it's impossible to visit a public restroom without knowing if someone has been chomping on this seasonal vegetable.

WHAT DO I NEED?
* cooked asparagus (boiled, salted asparagus is fine, but I recommend my favorite recipe at the end of this experiment, which adds immeasurably to the pleasure of performing it)
* a human urinary system
* a toilet

WHAT DO I DO? Cook the asparagus using your preferred recipe, eat it, and wait until you need to urinate.

WHAT WILL I SEE? Not much, but you may be aware, when urinating, of a peculiarly pungent odor.

WHAT'S GOING ON? The cause of the odor has perplexed biologists for a long time. While sulfurous compounds are known to be implicated, recent studies seem to suggest that it is in fact a cocktail effect that produces the unmistakable whiff. The combination of methanethiol, dimethyl disulfide, and dimethyl sulfone is the likely candidate. These substances are probably the result of the body breaking down the S-methylmethionine and asparagusic acid found in asparagus.

The difficulty in determining exactly which substances in what combination are responsible is compounded by the many variations in human production and perception of the odor. Several studies have indicated that production of the odor is a genetically determined trait exhibited by a maximum of 50 percent of adults. However, at least one report suggests that while all people produce the odor, only a minority are genetically able to smell it. So it may be that different individuals not only produce a different array of compounds but also have a different ability to smell them.

The reason that urine begins to smell so quickly after consuming asparagus may be due to the rapid response of the kidneys to any unusual products, such as the sulfurous compounds mentioned above. The body will quickly process and remove anything it considers potentially alien, and the asparagus odor can often be detected in urine within fifteen minutes of consuming the vegetable. Not that anybody should be alarmed—asparagus is an excellent, healthy food, it's just that your body removes the compounds it has no use for as quickly as possible, while simultaneously absorbing the beneficial ones.

P.S. Not everyone finds the smell offensive. Dr. Juvenal Urbino, a character in Gabriel García Márquez's *Love in the Time of Cholera*, "enjoyed the immediate pleasure of smelling a secret garden in his urine that had been purified by lukewarm asparagus."

My recipe for grilled, honeyed asparagus:
Parboil a bunch of asparagus spears—preferably young, tender tips—for five minutes. Meanwhile, combine two tablespoons of honey with one tablespoon of extra-virgin olive oil and mix well. Place the parboiled asparagus on a hot grill or under a broiler, and cook until the spears just begin to char. Brush them with the honey/oil mixture and grill until the honey begins to caramelize. Serve hot, sprinkled with sea salt. If you are feeling particularly sophisticated, combine eating the asparagus with quaffing the champagne from the "Fizz Fallacy" experiment in chapter 1.

☉ WINE INTO WATER

No matter what color a drink is, when the liquid finally leaves your body the color has gone. What happens to it?

Here's one that's perfect for kids. Ply them with cola, cranberry juice, or milk and then set them thinking about why the ultimate product is always yellow.

WHAT DO I NEED?
- a variety of drinks of different colors: red wine (for the adults), cola, water, cranberry or orange juice, milk, coffee, or anything else you can think of with an interesting hue
- a toilet
- a human digestive and urinary system

WHAT DO I DO? Sample the drinks in turn after an initial visit to the toilet to ensure that the next visit's urine will contain mainly the products of your most recent drink. You may have to do this over a period of days in order to establish a rhythm, perhaps consuming each drink at the same time on consecutive days. Make sure that you don't have anything else to drink between consuming each liquid and visiting the toilet. Note the color of your urine after each drink. If you are feeling adventurous after trying out the drinks from the list above, read below to see what happens if you eat too much borscht.

WHAT WILL I SEE? You may notice variations in the yellow color of your urine, but you will not, under normal circumstances, notice any coloring showing through from the original drink.

WHAT'S GOING ON? Colored substances in drinks (or food) are usually organic compounds that the human body has an amazing ability to metabolize, turning them into colorless carbon dioxide, water, and urea. The toughest stuff is taken care of by the liver, which is a veritable living waste incinerator, while the kidneys take care of removing waste products from your blood. By the time it leaves your body, the liquid is almost unrelated in chemical composition to the original liquid you consumed.

Any substance, solid or liquid, that goes down your esophagus and passes through the digestive tract, if not absorbed, is incorporated into fecal matter. Urine, by contrast, is produced by the kidneys from metabolic waste produced in the tissues that has been transported through the bloodstream. This waste is added to any excess water you have consumed to produce urine of various shades of yellow. This is stored in and passed out from the bladder.

Any colored compound that you drink either will or will not interact biochemically with the body's systems. If it does, this interaction (like any other chemical reaction it might undergo) will tend to alter or eliminate its color. If it does not, the digestive system will usually decline to absorb it and it will be excreted in the feces, which, you will have noticed, show considerably more variation in color than urine.

Some colored substances can make it through your digestive system and into your urine. This occurs when the intake of colored substances exceeds what the body can quickly metabolize, and the coloring is not removed as the liquid leaves the body. Anybody who wants to see this effect should consume a large quantity of borscht (beet soup). You'll notice that your urine takes on a distinctly pinkish hue. (If you are one of those rare people with highly acidic stomach conditions, then you may find the pigment is broken down more rapidly. Nonetheless, if you look carefully enough, there should still be a slight pinkish hue to the color of your urine.)

P.S. There is a current fad for drinking water in order to purify ("detox") the system after an excess of anything from alcohol to red meat to Thanksgiving dinner. But evidence suggests this is unnecessary. The minimum volume of urine required by the kidneys to excrete the waste products of the human metabolism is about half a quart a day. Because we lose another half quart through breathing, sweating, and defecating, we need to replace the total lost by drinking about a quart of water a day.

Drinking more than this merely dilutes the urine—the same amount of waste products (or "toxins" as they are popularly known) are still removed from our bodies. Coffee and tea have a mild diuretic effect, but your body still gets a net fluid gain from drinking them—you do not urinate more liquid than you consume. Alcohol, on the other hand, is a much stronger diuretic, so you should always drink lots of soft drinks or water after overindulging. "Detoxification" may be popular, but there is no evidence to support the argument that our kidneys cannot cope with the input and output of normal or even excessive consumption, as long as the latter is not on a long-term basis. Drinking more water than we need makes no significant difference to the elimination of waste products from our bodies. And indeed, it is possible to drink too much water—dangerously so.

☽ A LONG MEAL

So now that we know what can and can't pass through our bladders, what about our bowels? How long does it take to digest different types of food?

Again, children will love this one—the gross-out factor looms large. And for the really brave, there can be fun with a kebab skewer afterward!

WHAT DO I NEED?
- sugar
- bacon
- corn
- tomato

- mushrooms
- celery
- beets
- a human digestive system
- a toilet
- a stick or kebab skewer for poking through the products

WHAT DO I DO? Cook and eat the foodstuffs, plus anything else you fancy—fiber-rich vegetables produce the most interesting results.

WHAT WILL I SEE? Possibly nothing, depending on how closely you are prepared to inspect your feces. Plant material such as a red tomato is often visible in human excrement, as are the skins of peppers and capsicums. Corn is probably the most obvious retained product of the human digestive system, its yellow color standing out particularly well in feces. Celery and mushrooms are also difficult to break down, while beets can give a fascinating reddish hue, as it does to your urine.

WHAT'S GOING ON? Simple sugars such as glucose will be absorbed into the bloodstream relatively quickly and will not appear in any noticeable form after passing through the body. They may even pass through your cheeks and gums, because their molecules are small enough to pass directly into human cells.

However, most food needs to be processed by our digestive system before it will pass into our bloodstream and become useful fuel. The journey begins in our mouth, where the food is crushed until the pieces are small enough to swallow, and enzymes begin to break it down further. Once in the stomach, it is attacked by more enzymes and acids in gastric juices, and is blended by the churning motion of the muscles in the stomach wall. A couple of hours later this semi-liquid passes into the small intestine, where much of it is absorbed into the bloodstream.

Some components, such as dietary fiber, cannot be digested, and they move on to the large intestine. This takes about six hours. Once there, more water is absorbed by the body so that only the indigestible material remains. It may take up to thirty-six hours before the remaining waste is expelled through the anus. The final fecal form is approximately 75 percent water. The rest is bacteria, excreted biliary compounds, and unfermented fiber.

Of the various foodstuffs we eat, sugar is the first to be absorbed, followed by proteins (from eggs, nuts, and nonfatty meat) in about six hours, and finally by various types of fat over longer periods. The age, health, and size of the eater all affect the amount of time it takes.

As you will discover from carrying out the experiment, some components of food, such as fiber, are hardly broken down at all and, as in the case of corn, pass out of the body relatively untouched. The lignins in mushrooms and the cellulose in celery are also relatively unscathed after a journey through your body.

Those adults brave enough—and you can guarantee children will be keen—can, conditions and consistency willing, take a long kebab skewer and poke through the poo to see what they can recognize. If you do feel up to attempting this, disposable gloves are a sensible precaution, and you should certainly wash your hands thoroughly after disposing of the experimental material.

P.S. *New Scientist* readers are an ingenious bunch. In answer to a related question on how much excrement a human produces for a given amount of food eaten, they came up with the following information.

How much excrement you produce depends not only on how much food you have eaten, but also the type of food and the activity of the bowel. If you eat lots of high-fiber foods such as vegetables, beans, and cereals, which, as we have seen above, the body does not completely digest and absorb, you will produce more feces than if you eat lots of easily digested, low-fiber foods such as chocolate. Spicy foods, laxatives, and infections can affect the activity of the bowel. And the greater the speed of transit, the less water the gut can absorb and the greater the weight of feces produced, as anybody who has had gastroenteritis will testify.

The weight of feces produced by a healthy individual ranges from 19 to 280 grams (two-thirds of an ounce to nearly 10 pounds) per day. The only way to increase fecal weight is to eat more fiber, because unfermented fiber is indigestible and can hold lots of water. In healthy people, the wet fecal weight is in the order of 3 to 5 grams (0.1 to 0.17 ounces) per 1 gram (0.04 ounces) of fiber. From this we are delighted to be able to extrapolate the following formula for the effect of fiber in the colon:

$$\text{Feces weight} = W_f(1 + H_f) + W_b(1 + H_b) + W_m(1 + H_m)$$

where W_f, W_b, and W_m are respectively the dry weights of fiber remaining after fermentation in the colon, bacteria present in the feces, and osmotically active metabolites and other substances in the colon that could reduce the amount of free water absorbed. H_f, H_b, and H_m are constants denoting their respective water-holding capacities.

WANT TO READ MORE? For the most serious students, www.constipationadvice.co.uk/constipation/digestive_system.html.

◯ WAYWARD WATER

Why does an electrostatically charged plastic comb attract a stream of water?

Don't be concerned by the terminology. To charge your comb electrostatically, all you need to do is comb your hair with it.

WHAT DO I NEED?
- a plastic comb (not a metal one)
- some hair
- running water
- molasses

WHAT DO I DO? Comb your hair, or someone else's if, like me, you don't have any. Turn on your tap until you get a thin, smooth, steady flow of water. The thinner the stream, the better, as long as it does not break down into droplets. Then place the comb near the point where the water emerges from the tap. Afterward, try the same experiment with molasses. (You might also want to compare the effect on display here with the one from the "Shape Shifting" experiment in chapter 1.)

WHAT WILL I SEE? The water is attracted to the side of the flow where you are holding the comb. Large deflections of the water stream can occur.

WHAT'S GOING ON? The molecules of water from the tap are electrically neutral overall but, like the molecules of other dielectric substances, they are characterized by regions of positive and negative charge. The electrical field caused by the comb (which is produced by an excess of electrons collected by the comb from the friction caused as it runs through your hair) attracts the more positive parts of each water molecule and repels the negatively charged regions.

The water molecules are rotated and stretched by the field and tend to become aligned head-to-tail along the electrical field lines, with their positive ends pointing toward the comb. The charges on each molecule therefore neutralize each other except at the water/air interface that is closest to the comb, and also at the interface furthest from it. In these regions they are not paired with opposite charges (because they are at the edge of the water stream and there are no molecules on one side of them), so they build up areas of equal positive and negative charge on the opposing sides of the cylindrical water surface. The stream is now polarized, and it is this property that allows dielectric materials to be attracted to a source of electrical field—in this case the comb.

Comb your hair.

The comb is now negatively charged with electrons.

Hold the comb near a slow, thin flow of water from a tap.

The electrons in the comb attract positive charges in the water molecules (and push away negative charges).

Water is not unique in showing what is called the Doff effect. Try the same experiment with molasses. Dribble a thin filament of molasses from a tablespoon. Because the molasses is more viscous than water and therefore flows more slowly, the electrical field from the comb has more time to act on each section of the flow's length. The resulting deflection is spectacular.

9 SPECTRAL IMAGES

When you draw pictures in condensation on a bathroom mirror, they disappear when the condensation evaporates, but return when the room steams up again. Why?

A great way to demonstrate this phenomenon to children—without risking that some rude words will reappear on your mirror when Great-aunt Gladys takes a shower—is to challenge them to a game of tic-tac-toe. If they win, they can revisit the victory at their next bath time.

WHAT DO I NEED?
* a bathroom that readily steams up
* a mirror
* a finger

WHAT DO I DO? Wait until you or the kids are taking a hot bath or shower (there's no good reason to waste water or heating) and your bathroom has steamed up, then draw or write on your mirror with your finger. After you've finished bathing, open the windows and let the room cool. The next time you or a family member takes a bath or shower and the room steams up, look at the mirror.

WHAT WILL I SEE? Your original message or picture, which vanished when the room cooled down and the condensation on the mirror evaporated, will return when the bathroom steams up and the mirror mists over again.

WHAT'S GOING ON? When water vapor condenses on a dry mirror, it does so as separate droplets—a process known as dropwise condensation. These droplets effectively screen the mirror so that it appears opaque.

When you draw on the surface with your finger, the droplets coalesce into a thin film of transparent water, so the mirror becomes reflective again in these areas—this is called filmwise condensation. When the mirror warms up, or the air humidity falls, the droplets evaporate and the image disappears because the surrounding droplets no longer contrast with it.

However, the film of water evaporates more slowly than the droplets because of its lower surface area. If it does not have time to evaporate completely before the bathroom steams up again, any condensation occurring soon afterward will be dropwise where there were droplets before and filmwise where some of the film remains. The image will then reappear on the glass.

This only explains how mirror drawings return in the short term. If the mirror dries completely, the pattern should not normally reappear when further condensation occurs. But this presumes that your drawing finger is entirely clean, which is unlikely. When you draw an image in the condensation, your finger will almost certainly leave behind traces of grease or sweat, plus possibly shampoo or soap. These traces are transparent, so when the condensation disappears you can't see them. The next time water vapor condenses on the mirror, however, there is a noticeable difference in droplet size between those forming on the clean glass and those forming on the greasy or soap-contaminated glass.

Grease will tend to repel water droplets while water-loving surfactants such as soap will reduce the droplet size and generate a smoother, clear film of water. Whatever the cause, as the mirror steams up again, the image you drew will contrast with the opaque mist on the surrounding glass.

P.S. Dropwise condensation is known by chemical engineers to be more efficient at transferring heat than filmwise condensation, but in practice it is much more difficult to promote, because as the droplets enlarge, they touch each other and coalesce, so the process tends to become filmwise.

On the other hand, dropwise condensation is easy to prevent. Wiping the mirror with a cloth or a tissue wetted with a small amount of detergent such as shampoo leaves an invisible film on the surface, as we have seen. This reduces the surface tension of the condensing droplets, causing them to flatten out and readily coalesce into a film. This is the basis of antimisting fluids used for treating eyeglass lenses and the inside of car windshields.

○ BRUSH MUSH

Why does orange juice taste so awful after brushing your teeth?

You know how it is. You got up late and you're in a rush. You swig a glass of orange juice right after you've brushed your teeth. Instead of a sweet fruit flavor, the juice tastes like a foul bitter concoction. At its worst, the rest of your breakfast can taste revolting, too. Something's going on. . . .

WHAT DO I NEED?
◆ orange juice
◆ three types of toothpaste: ordinary mint; mint without foaming agents (available at stores stocking organic products; check the label for the absence of sodium lauryl sulfate [SLS], which is a foaming agent used in many toothpastes); nonmint (again, one that doesn't contain SLS)

If you can't get hold of the SLS-free varieties of toothpaste, you can still try the experiment using the ordinary mint toothpaste.

WHAT DO I DO? Take a swig of orange juice and note how sweet it tastes. Brush your teeth with nonmint toothpaste, then take another swig of juice and see how it tastes. Repeat with the SLS-free mint toothpaste and then with the standard mint toothpaste.

WHAT WILL I TASTE? After brushing with nonmint, nonfoaming toothpaste, you might expect the juice to taste foul, but surprisingly, it tastes just like normal sweet juice. Brushing with nonfoaming mint toothpaste makes the juice taste weird, but not too bad. However, normal mint toothpaste transforms the juice into a bitter, medicine-like flavor and strips all the sweetness away.

WHAT'S GOING ON? You've just experienced the "orange juice effect," something that's been known about since the 1970s. Foaming agents, usually sodium lauryl sulfate, are added to toothpaste to help disperse the paste around your mouth, and for easy rinsing, but they have a nasty aftereffect. They interfere with the taste buds on your tongue, suppressing the ability to taste sweetness and salt and enhancing any bitter flavors. The powerful mint flavor that most of us want from our toothpaste simply adds to the problem. While it doesn't interfere with our taste buds in the same way as the foaming agents, the strong mint flavor overpowers the taste of anything that is consumed afterward, including the breakfast you were looking forward to. The rule—and it makes sense for the health of your teeth, too—is brush your teeth after meals, not before.

P.S. Foaming agents aren't a vital ingredient, but, as with soap, if we don't see any bubbles we think it isn't working. The same goes for mint. If a toothpaste doesn't knock you out with its mint flavor, we assume it's not effective. In Japan, most toothpastes are fruit flavored, and in the West many children's toothpastes are fruity, not minty, too.

6   IN THE GARAGE (OR THE UTILITY ROOM . . .)

◯ HOT STUFF

Does hot water really freeze faster than cold water?

Astoundingly, yes it does. And not only that but *New Scientist* magazine can lay claim to being the first to publish this discovery in the modern era when a Tanzanian student, Erasto Mpemba, wrote to the magazine in 1969, pointing out that when he was trying to make ice cream he found that the mixture froze more quickly when put in the freezer warm than if allowed to cool to room temperature first. Readers were skeptical, but experiments with both ice cream and water—distilled or tap—proved that Erasto had spotted a quite extraordinary phenomenon. But why does it happen?

WHAT DO I NEED?
- flattish plastic containers or ice cube trays
- warm water at 95°F (35°C)
- cold water at 41°F (5°C)
- a freezer

Or:
- two metal buckets

- warm water at 95°F
- cold water at 41°F
- freezing night air

You'll need a thermometer to measure the water temperature if you want to be truly accurate, but if you don't have one, you can use water at around these temperatures—one close to body temperature and the other a little above freezing.

WHAT DO I DO? Fill one container or tray with the warm water, a second with the cool water, and place them in the freezer. Give them at least ten minutes before checking them, then check again at regular intervals. If you are using the buckets and a cold night, you'll need to fill them and wait some time longer.

WHAT WILL I SEE? The warm water will freeze faster than the cold water. It really will! The only limitation in the case of the freezer experiment is that the water containers must be relatively small so the capacity of the freezer to conduct away the heat content is not a limiting factor.

WHAT'S GOING ON? Initially it was believed that placing the containers on a freezer shelf allowed the container with the warmer water to melt the icy surface on which it was resting so that it settled into a nice frosty resting place. That meant more of the container was surrounded by ice and therefore it had greater contact with a cold surface than the cooler water container. The increased rate of heat transfer from the warmer container and its contents therefore more than offset the greater amount of heat that had to be removed.

However, while this is clearly an accelerating factor, further research showed that the effect occurred even if the containers were sitting on a dry surface or were suspended above the freezer shelf.

It was then suggested that a reduction in the volume of the warmer water due to evaporation might make that body of water smaller and so cool more quickly, but this is not the case. Thermometers placed in the water show that the cooler water reaches the freezing point of 32°F (0°C) more quickly than the warmer water, as you would imagine and as predicted by Newton's law of cooling. After that, however, the water that had started warmer solidified more quickly. In fact, in experiments, the maximum time taken for water to solidify in the freezer occurred with an initial temperature of 41°F (5°C) and the shortest time at about 95°F (35°C), which is why we suggest those temperatures above.

This seems paradoxical, but can be explained by understanding that there is a vertical temperature gradient in the water. The rate of heat loss from the upper surface is proportional to the temperature of the surface. So, if the surface can be kept at a higher temperature than the bulk of the remaining liquid (in this case, the initially warmer water), then the rate of heat loss will be greater than from liquid with the same average temperature uniformly distributed (in this case the initially cooler water). This can be proved by using

tall, thin metal cans as water containers. The paradoxical effect disappears in this case because the temperature gradients in the cans are lost as heat is rapidly
conducted through the metal sides much faster.

Nonetheless, there may be other factors at work.
Cold water forms its first ice as a floating skin, which
impedes further convective heat transfer to the surface.
Hot water, on the other hand, forms ice over the sides
and bottom of the container, while the surface remains
liquid and relatively hot, allowing radiant heat loss to
continue at a faster rate. The large temperature difference drives a vigorous convective circulation that
continues to pump heat to the surface even after most
of the water has become frozen.

This argument is further borne out when conducting the experiment using metal buckets full of
water in open air on a cold, preferably windy, night
(which promotes heat loss). If the initial temperature
of the water is around 41°F, cooling of the core is very
slow, particularly as loose ice floats to the surface,
inhibiting normal convection. There is no means by
which the slightly warmer water in the bucket can
come into contact with the cold bucket edges and
transfer its energy to the outside.

But if the initial temperature of the water in the
bucket is around 95°F, strong convection is established
before any water freezes, and the entire mass cools
rapidly and uniformly. Even though the first ice will
form later than in the cool water bucket, complete
solidification of the water in the warmer bucket can
occur more quickly than if the water starts off cool.

But those temperature differences are critical. Obviously, if the cold bucket starts very close to freezing, say at 33°F or 1°C, and the hot bucket starts very close to boiling, say at 211°F or 99°C, the experiment is unlikely to cause surprise. Additionally, the containers must be large enough to sustain convection with a small temperature gradient, but small enough to extract heat quickly from their surfaces. This is where the forced cooling of a windy night helps.

And there's more: recent, as yet uncorroborated, research from Washington University, in St. Louis, has offered yet another possibility. Solutes, such as calcium and magnesium bicarbonate, precipitate out if water is heated. These can be seen lining the inside of any kettle used to boil hard water. However, water that has not been heated still contains these solutes. As it freezes, the ice crystals that form expel the solutes into the surrounding water. As their concentration increases in the water that has yet to solidify, they lower its freezing point in the same way salt sprinkled on a road in winter does. This water therefore has to cool further before it freezes. Additionally, because the lowering of the freezing point reduces the temperature difference between the liquid and its surroundings, the heat loss from the water is far less rapid. Nonetheless, the fact that both tap water and distilled water (the latter being virtually free of solutes) both display the effect described in this experiment shows that this is not the decisive factor.

Finally, there could be yet another factor affecting this process—supercooling. Research shows that because water may freeze at a variety of temperatures, hot water may begin freezing before it is entirely cold. But whether it will completely freeze first may be a different matter.

Just one more issue remains, but it involves a technological rather than a physical dimension. The temperature oscillation inside a freezer depends on the sensitivity of the thermoelement and the timer that controls the system. We may assume that at the freezer's standard temperature, the power used for cooling the freezer operates at a standard rate. So if a bucket of cold water is added, it may produce only a small effect on this power output because it will not trigger the temperature sensor. However, a bucket of hot water may easily activate the sensor and release a short but powerful cooling of the freezer with an additional cooling overshoot, depending on the timer. While this would not account for the version of the experiment conducted outdoors on a cold night, it might explain some of the freezing differences seen in the domestic freezer version of the trial. This would only apply, however, if the experiment was being carried out with the hot and cold containers being placed in the freezer independently of each other in two separate experiments. By placing them in the freezer together, as suggested above, the variable can be excluded.

It seems that a number of factors are at work here, with no one theory acting in isolation. Do take time to try out any number of combinations of temperature, vessels, and environments—after all, fame awaits whoever stumbles on the ultimate answer.

P.S. This effect was noticed by Sir Francis Bacon in 1620 using wooden pails of water standing on ice. Even earlier, in 300 BC, Aristotle noted that hot water froze faster than cool water. This is Aristotle's account from *Meteorology:* "Many people, when they want to cool water quickly, begin by putting it in the sun. So the inhabitants when they encamp on the ice to fish (they cut a hole in the ice and then fish) pour warm water round their rods that it may freeze the quicker; for they use ice like lead to fix the rods."

Nonetheless, and despite the phenomenon's illustrious history, it was Erasto Mpemba who dragged the paradox into our era. If anybody knows what became of him or where he lives now, we'd be delighted to hear from him.

9 FREEZER TEASER

How can you grow spikes of ice in your freezer?

If you are tired of mini-umbrellas and olives on cocktail sticks in your drinks, cubes of ice with dangerous-looking spikes sticking out of them might just enliven your parties.

WHAT DO I NEED?
- a domestic freezer (preferably a frost-free model)
- two plastic ice-cube trays
- tap water
- distilled water

WHAT DO I DO? Fill one tray with distilled water, the other with tap water, and place both in the freezer. If your freezer has a temperature control, it seems around 19°F (−7°C) is the ideal spike-growing environment.

WHAT WILL I SEE? Ice spikes should begin to grow from the surface of the ice-cube tray containing distilled water. They are less likely to grow out of the control tray, containing tap water, but they might. Sometimes no spikes will form in either tray, so repeat the experiment until you see what you are looking for.

WHAT'S GOING ON? First, a bit of history. Ice spikes have been noted growing out of frozen bodies of water for some time, but nobody knew why they formed only occasionally. Then, in 2003, Ken Libbrecht and Kevin Lui from the physics department at the California Institute of Technology in Pasadena began to look into the mystery.

Libbrecht had been growing designer snowflakes in his laboratory to find out why they formed such complex and delicate patterns and was used to creating specialist ice formations. When he was sent photographs of spikes protruding from a tray of ice cubes, he was intrigued and started trying to grow them himself in his home freezer. Not surprisingly, most of the cubes he froze turned out, well, cubic, but once in a while a spike would grow out of one of them. When he began trying the experiment with different compositions of water, he discovered that distilled water produced more spikes per ice-cube tray than tap water.

Why? After about ninety minutes in the freezer the water surface on the ice cubes starts to freeze, initially around the edges of the cube. This is because the plastic tray is covered with microscopic nicks and scratches and water molecules can wedge themselves into these tiny hollows. Here the molecules can form plenty of bonds with their neighbors as the temperature falls without being moved around in any convection currents occurring in the bigger body of water. Because ice crystals are less dense than water, they float to the surface. The freezing ice then creeps slowly toward the

Fill the tray with water and place it in the freezer. Ice forms first at the edges and on the surface, leaving a hole in the middle.

water

The ice thickens and pushes water up through the hole where it freezes.

ice spike

Eventually the tip freezes over, creating the ice spike.

middle of the water surface until only a small hole remains unfrozen at the center of the ice cap.

At the same time, more ice starts forming around the sides and bottom of the cube. Because ice expands as it freezes, the ice growing below the surface pushes water up through the hole. If the conditions are just right, the meniscus of water forced out of the hole freezes around its rim, forming the base of a hollow spike. As this process continues, the spike grows taller until all the water has traveled up its hollow interior and frozen or, more commonly, the tip of the tube freezes over. Interestingly, spikes reach their full height in no more than ten minutes.

The reason that tap water produces fewer spikes than distilled water is due to the impurities in the former. As the tap water freezes around the top of a growing tube of ice, its saltiness increases, because dissolved minerals and metals remain in solution as long as unfrozen water is present. They are therefore contained in the last bit of the water to freeze. They build up to such high levels at the tip of the spike that the water there can no longer freeze and the growth stalls. So to get the best ice spikes use distilled water—the kind sold for car batteries or steam irons.

Libbrecht and Lui have more advice for those wishing to grow ice spikes. Temperature makes a big difference. At 19°F (−7°C), about half of distilled water ice cubes turn spiky, but only one in ten shows a spike at 5°F (−15°C). This may be because the tips of the embryonic ice spikes freeze over before they have a chance to grow. A frost-free freezer helps, because in these models a fan is continually circulating cold, dry air. The moving air chills the edges of the water droplet perched atop a growing ice spike faster than in calm conditions, and this promotes the growth of longer ice spikes. This works even better if you allow the air to circulate around your freezer by removing all those pizza boxes and frozen chicken wings. Finally, plastic ice-cube trays work far better than metal ones. Metal conducts heat away from the cubes so quickly that the water surface freezes and seals the hole from which the spike would normally grow before it gets a chance.

P.S. It's possible to grow ice spikes in your backyard, and these can be even more impressive than those grown in your freezer. You need a small container such as a plastic saucer or bowl, which you should fill to just below the rim—again, you'll have better results with distilled water. You'll also have to wait for a freezing night that, as in the case of the frost-free freezer, would be made more ideal by the presence of a light breeze. If you are fortunate, you'll get an ice tower taller than those seen in the freezer and, if the breeze is light and constant but not strong, the tower will probably lean in one direction. It may even be triangular in cross section as the wind flows around the forming tower. An outdoors ice spike grown in New Zealand in July 1991 formed a helical, screwlike icicle with eight complete turns as the breeze changed direction in swirling conditions.

It seems the record length for a freezer spike grown from an ice-cube tray is 2.2 inches, but those grown from double-distilled water in larger containers can reach four inches. Ice spikes grown outside can exceed nine and a half inches in height. If you find a fool-proof way of beating these records, we'd love to hear from you.

WANT TO READ MORE? The December 20/27, 2003–January 3, 2004, issue of *New Scientist* (page 38) has full details on the discovery and creation of ice spikes.

⊙ CUBED ROUTE

Why is the highest concentration of bubbles in ice cubes always found in the middle of the cube?

The answer to this question is closely related to the formation of ice spikes we saw on page 154 in the "Freezer Teaser" experiment. To prove (a) that scientists are a lazy bunch and won't do any work beyond that which is necessary and (b) why the bubbles form in the pattern noted above, you can prepare this experiment at the same time as growing your ice spikes and run both alongside each other.

WHAT DO I NEED?
- a domestic freezer
- an ice-cube tray
- tap water

WHAT DO I DO? Fill the tray with tap water and place it in the freezer. If you are running the "Freezer

Teaser" ice spike trial alongside this one, you'll want to set the freezer temperature control (if you have one) to 19°F ($-7°C$), although it's not necessary to set a specific temperature for the purposes of this particular experiment.

WHAT WILL I SEE? When the cubes are frozen, you'll notice that in most cases there is a concentration of bubbles in the center of each cube, with spines of progressively smaller bubbles radiating outward.

WHAT'S GOING ON? Two effects covering the solidification of ice are at work here: the segregation of a solution into two separate physical forms or phases as it solidifies, and the way in which ice crystals tend to grow.

Tap water contains on average 0.003 percent dissolved air by weight, but the solubility of air in ice is very small, so the first crystals that start to grow after you have placed your ice-cube tray in the freezer contain almost no dissolved air. As the tray cools, heat flows away from the water through each of the outer faces of the cube and the first crystals start to form on the outer walls of the cubes, growing inward as more ice appears in the form of columnar crystals. These early, outer crystals are clear, because they contain no dissolved air.

However, the unfrozen water, which still lies between these early, airless crystals in the center of the cube, gradually becomes saturated with the air that is being driven out of those first crystals. When the concentration of air in this remaining water reaches 0.0038 percent, it forms a peritectic composition—a mixture that has the lowest freezing point of all possible mixtures of that particular substance. When this occurs, all the remaining liquid freezes, separating into a mixture of approximately 2.92 percent air by volume and 97.08 percent ice by volume.

The smaller bubbles radiating outward in the frozen cubes are from the peritectic mix that was trapped between the early, clear columnar crystals. The high concentration of bubbles in the middle is where the remaining bulk of the peritectic mix had been trapped. Hence the traditional ice-cube bubble pattern.

Now pop the cubes you've just made into the James Bond vodka martini left over from the "You Only Drink Twice" experiment in chapter 3 and sit back knowing more about the science of enjoying yourself than you did before.

P.S. An extreme example of the formation of columnar crystals can be seen if you bite into a popsicle. You'll see crystals radiating inward from the outside of the popsicle toward the center. These were created as they grew inward from the mold the popsicle was made in. You might also see columnar crystals and the segregation of dissolved elements (in the same way the bubbles form in ice cubes) in cast metals and candles.

Incidentally, it is very difficult to create bubble-free ice cubes; the ones you see in TV ads are usually fakes, made from glass or Perspex so they won't melt under the studio lights. However, if you want to try to make some at home it can be done. Use boiling water to remove some of the dissolved gases. Leave the boiled water to cool in its container, covered with plastic wrap to prevent air from redissolving in it. Now freeze the water very slowly so that there is no steep temperature gradient and the bubbles, when they appear, do not get trapped but have time to move through the liquid and up to the surface before freezing into place.

Commercial ice-cube machines create clear cubes by freezing a small portion of water at a time and then trickling more water over the ice cube's surface so that the gas-carrying component runs off while layer after layer of non-gassy ice builds up.

9 CLOUD BURST

How can clouds be created in a plastic bottle?

Even more interesting, you can make them vanish and reappear just by squeezing the bottle.

WHAT DO I NEED?
- water
- a clear, flexible two-liter plastic bottle with a screw cap
- matches

WHAT DO I DO? Add just enough water to the bottle to cover the base, and shake it around. Light a match, let it burn for a couple of seconds, then blow it out. Immediately drop the smoking match into the bottle and screw the lid on quickly and tightly. Now squeeze the bottle hard four or five times.

WHAT WILL I SEE? A cloud will appear in the bottle after you have added the match. But when you squeeze the bottle the cloud disappears. Release the bottle, and there is the cloud again.

WHAT'S GOING ON? In effect, you have built a small cloud chamber. The cloud in the bottle forms as a result of three factors: the smoke from the match, tiny drops of water inside the bottle that were formed when you shook the bottle with the water added, and the change in air pressure caused by squeezing or releasing the bottle.

Water in the bottle combines with air to form water vapor—the gaseous form of water. This gives rise to the cloud in the bottle. But for water vapor to form this cloud, particulates also need to be present. In this case, smoke provides the necessary particles that act as nucleation sites, allowing drops to collect on them. Without the smoke, no cloud will appear.

Squeezing the bottle raises the pressure. This increases the temperature inside the bottle, helping the water to change from visible liquid back to invisible gas (most liquids turn into gases as temperature increases), and the cloud disappears. Releasing the pressure reverses the effect.

P.S. Real clouds form in exactly the same way. Water vapor comes from the evaporation of seawater, rivers, and lakes. This expands and cools as it rises. There is a limit to the amount of water vapor the air can hold, and this limit is higher when the air is warmer. As the temperature drops at higher altitudes, the vapor begins to condense and clouds are formed, just as in the bottle. At higher altitudes the gas expands, is less compressed, and is cooler, so clouds appear, as they do when the bottle is left alone. At lower altitudes the gas is warmer and more compressed, so there are no clouds, just like when the bottle is squeezed. And, because the atmosphere contains many small particles, ranging from dust to smoke to salt particles, there are plenty of nuclei around which condensation can gather to form clouds if the pressure and temperature conditions are correct.

○ CLEANING UP

Why is cola so good for cleaning dirty coins?

It's not only dirty money that you can launder with cola; brass and similar alloys can be brought to sparkling luster using your favorite soda.

WHAT DO I NEED?
- a large glass or bowl
- a few cans of cola
- a handful of loose change (preferably copper coins, such as pennies minted before 1982)

If you are happy with the results of cola on your coins, you might also try this technique out on household objects made out of brass or other alloys, such as doorknobs and screws.

WHAT DO I DO? Fill the glass or bowl with cola and drop in your coins.

WHAT WILL I SEE? Leave the coins for a few hours while you try out one or two of the other experiments in this book. When you return you'll notice how bright and shiny your once-tarnished coins are. If they are particularly dirty you may need to give them a rub with a tissue or cloth, but when you do you'll be pleased to see they look as though they were minted yesterday.

WHAT'S GOING ON? The tarnish that forms on an alloy is caused by a reaction between oxygen and one of the metals that make up the alloy. This forms an oxide. Cola, which is weakly acidic, reacts with the oxide to form a soluble salt of the metal that produced the tarnish, which then dissolves in the liquid, leaving behind a clean metal surface as bright and shiny as the day the coin was minted.

Cola contains phosphoric acid. Because this acid can eat away at substances other than the tarnish on coins, you should always brush your teeth after drinking cola. And because regular cola and diet cola both have the same cleaning effect, don't be fooled into thinking that diet cola may in some way be better for your teeth than regular cola. It does have less sugar, but the phosphoric acid levels are very similar.

The citric acid in lemons and the acetic acid in vinegar are also excellent for removing the oxide on copper coins. If you clean a lot of copper coins in vinegar, you'll notice the liquid takes on a green hue, which is the copper acetate formed during the reaction between copper oxide and acetic acid. Beware though: copper salts are poisonous, so don't consume the vinegar (or, indeed, the cola) afterward. And remember, while this method is great for cleaning your loose change, it should not be used for valuable collections of antique coins and the like.

P.S. *New Scientist* reader Rob Hay wrote in to assure us that even in light of the effect that cola has on coins, it has little effect on the stomach—despite its acidity—because the stomach is even more acidic than the cola. In fact, he has anecdotal evidence to show this. As a keen and sometimes thirsty cyclist in his youth, he often spilled soft drinks over the paintwork on his bike with no ill effects. But on one cycling trip he became rather ill and, as he was dismounting from his bike, he was violently sick. Ninety minutes later, the areas of the frame that had come into contact with his vomit had been completely stripped of paint. The contents of the stomach, it seems, are far more corrosive than the contents of a can of cola.

9 STICKY SOLUTION

How does molasses clean rusty objects?

Molasses rivals cola in its ability to buff up tarnished household objects. Which foodstuff you prefer to use probably depends on what it is that needs cleaning.

WHAT DO I NEED?
- molasses
- water
- a large plastic bucket
- some rusty household items (door hinges, bicycle chains, and old nuts and bolts all fit the bill)

WHAT DO I DO? Mix one part molasses to nine parts water in the plastic bucket. Completely submerge the rusty item in the liquid and leave it for two weeks.

WHAT WILL I SEE? The article will emerge clean and shiny from the mixture.

WHAT'S GOING ON? Molasses contains chelating agents. These are made of molecules shaped a bit like the claws of a crab—the word "chelation" comes from the Latin *chele*, meaning claw. These agents can envelop metal atoms on the surface of an object, trapping them and removing them. Molasses owes its properties to cyclic hydroxamic acids, which are powerful chelators of iron.

More of these compounds are found if the molasses is derived from sugar beets rather than cane sugar. The plants from which molasses is made presumably use these chelating agents to extract minerals from the soil. Interestingly, there are aerobic microorganisms that use similar cyclic hydroxamic acids to scavenge iron. So plants and microbes appear to use the same chelation strategy to obtain their daily ration of iron.

A similar process is at work when old coins are cleaned with cola. Phosphoric acid, a key ingredient in cola, is a chelating agent. The power of a chelating agent also explains why the insides of tomato cans are lacquered. The citric acid in the tomatoes would dissolve the metal of the can if the lacquer were not present. Household cleaning agents, especially detergents and shampoos, also rely on chelation. These soften water to make it more effective during the cleaning process.

Chelation has its uses in medicine, too. EDTA (ethylenediamine tetra-acetic acid) is used as a chelating agent to control calcium in the body and can reduce the effects of mercury or lead poisoning.

P.S. Brown sugar is the whole-wheat flour of the sugar world, and white sugar is the refined version. When white sugar is extracted from the juice of sugar cane, the dark, nonsugar substances left behind are known as molasses. Most of this material is used as animal feed, but molasses is also an excellent source of minerals and vitamins—a tablespoon of blackstrap molasses has an iron content similar to that of a ten-ounce sirloin steak, though it is still 70 percent sugar.

9 STICKING DOORS

Why does a freezer door that has just been opened and closed stick shut so tightly that it is almost impossible to reopen?

We've probably all noticed this. You take out the ice cream and shut the door only to realize that you've forgotten to take out tomorrow's chili con carne to thaw out. But the door is now resolutely stuck. . . .

WHAT DO I NEED?
- a freezer that has been on for at least twenty-four hours

WHAT DO I DO? Open the freezer door for a few seconds, then close it for a few seconds. Now try opening it again.

WHAT WILL I SEE? The second time you try to open the door it's much more difficult. In some cases you'll end up pulling the freezer toward you as you tug at the handle. In others, the door will just remain closed, and you'll have to wait before the effect wears off and you can open it again as easily as the first time.

WHAT'S GOING ON? Freezer doors stick because opening the door the first time allows some of the cold air to flow out of the bottom of the freezer. If you stand in bare feet at the door of an open freezer you will notice this cold, dense air chilling your toes. The cold air flowing out at the bottom allows warmer air at room temperature into the top of the freezer.

When the freezer is closed again, this new air cools and contracts, creating a partial vacuum and making the door seem to stick as it is pressed to the seal by the higher air pressure outside the freezer.

If the day is particularly warm, there is a greater temperature difference between the exterior room temperature and the interior freezer temperature. In this case it is sometimes necessary to pry open the door seal a little with your fingers to allow some air in to equalize the pressure inside and outside the freezer before you can get the door open.

The partial vacuum created soon passes, however, because the door seals are not absolutely airtight and the air pressure inside and outside the freezer is equalized by air leaking into the compartment.

It's worth pointing out that if you try this experiment and you don't notice the sticking-door effect, you should check the seal. It may be leaking a lot of cold air and consequently making your freezer work much harder than it should, to say nothing of the monetary and environmental costs involved.

P.S. The vacuum effect is one reason among many why it is very dangerous to let children play with fridges and freezers. Certainly they should be told never to climb inside an empty one. This experiment should be carried out only when an adult is present.

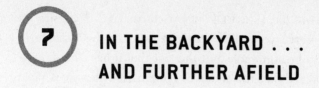
☾ FOSSIL RECORD

When it dies, I'd like my pet hamster to become a fossil. How can I make sure this happens?

Remember, whatever applies to a hamster applies to humans, too. So if you would like to become a fossil, pay attention to what follows, make detailed notes in your will, and find a sympathetic funeral director.

WHAT DO I NEED?
- a naturally deceased hamster (or other pet)
- a variety of environmental conditions

WHAT DO I DO? Take your hamster to an appropriate natural environment and ensure that it will not be removed from its final resting place by scavengers or natural phenomena.

WHAT WILL I SEE? Very little. Fossils take tens of thousands of years to form, but you will be saving up enjoyment for future generations of paleontologists.

WHAT IS (SHOULD BE) GOING ON? A desire to preserve your fluffy pet's remains in fossilized form may be admirable, but the fact that it has a hard, mineralized skeleton and lives a nonmarine lifestyle is a bad start. Terrestrial conditions are liable to erosion and significantly reduce the chance of fossilization. The soft tissue surrounding the skeleton of mammals decays very quickly—it is usually preserved only if the animal dies at high altitude or in a freezing environment, such as a glacial crevasse or in the polar regions, and then only in wizened, mummified form, which is not true fossilization.

So, if you really want your pet to survive the ravages of geological time, then while it is still alive you need to concentrate on improving the quality of its teeth and bones. Fossilization of these involves additional mineralization, but you can give your hamster a head start by feeding it a diet rich in calcium to build up its bones and teeth. If you are seriously thinking of becoming a fossil yourself, you should get a good dentist. If your hamster has eaten a few seeds in its last days, these, too, can become fossilized and intrigue paleontologists in the millennia to come.

After that it's a matter of location, location, location. You need to bury your dear departed rodent where it won't be disturbed for a very, very long time. Fossilized remains are often found in caves, so you could take up spelunking to scout out suitable locations (after proper training, of course). Alternatively, you need to find a spot where your hamster will be buried quickly after it has been laid to rest—preferably somewhere natural and dramatic, the sort of site from which you can detect a distant volcanic rumble and clouds of ash being emitted skyward. Don't get too close though: an ash burial is good, incineration by flowing lava is not. Again, be very careful; you don't want to join your hamster in fossilized harmony at this stage.

As the volcanic option suggests, you may have to travel a long distance to find the optimal conditions. A desert wadi in the flash-flood season offers a good environment, as does a tropical river floodplain during heavy rain, so you can bury your hamster in fine, anoxic (oxygen-free) mud.

However, the best environment to ensure a fossilized hamster is probably a sea burial in very deep water (shallow marine conditions are turbulent and full of organisms that will disturb or eat the remains). There are few creatures in the deep sea, and even fewer below the sediment and mud. As long as your hamster is not buried near a tectonic subduction zone where the earth's crust is being consumed, such as the Pacific Coast of the United States and the entire Americas, it should rest undisturbed until fossilization takes place.

This environment and the land-bound ones described above are ideal. The oxygen-free conditions will slow decay of the body, and the ash or fine seabed clay will help to preserve the body structure. Fossilization will then proceed until your hamster is nothing more than an outline of carbon and petrified body fluids, thanks to compaction from the weight of sediment that settles above. You should allow at least two hundred thousand years for this.

Of course, there is more chance of your winning the

lottery than you or your pet ending up as a fossil, but that's no reason not to try.

P.S. The following advice to let stalactites do the work of an embalmer is found in a poem written by Richard Whatley in 1820. The subject of the poem, William Buckland (1784–1856), was one of the most famous geologists of his day—and a noted eccentric, who claimed to have eaten his way through the entire animal kingdom. His contemporary, Augustus Hare, recorded how Buckland came upon a casket containing the preserved heart of a French king, to which Buckland exclaimed, "I have eaten many strange things, but I have never eaten the heart of a king before," and promptly gobbled it up.

Elegy Intended for Professor Buckland
Where shall we our great Professor inter,
That in peace may rest his bones?
If we hew him a rocky sepulchre,
He'll rise and break the stones,
And examine each stratum that lies around—
For he is quite in his element underground.
If with mattock and spade his body we lay
In the common alluvial soil,
He'll start up and snatch those tools away,
Of his own geological toil;
In a stratum so young the Professor disdains
That embedded should lie his organic remains.
Then exposed to the drip of some case-hardening spring,
His carcase let stalactite cover,
And to Oxford the petrified sage let us bring,

When he is incrusted all over;
There, 'mid mammoths and crocodiles, high on a shelf,
Let him stand as a monument raised to himself.

Finally, and it should go without saying, if you intend to attempt fossilization of your pet, wait until it has died of natural causes before carrying out the above instructions.

⑨ GOGGLE-EYED

Why can I see clearly underwater when I'm wearing swimming goggles or a mask, but everything looks blurred if I'm not?

Our eyes are adapted to viewing in air, but what goes wrong when you open them underwater without first putting on your goggles?

If, like me, you wear contact lenses, make sure you remove them before attempting the part of this experiment that requires you to open your goggle-free eyes underwater. It might make scientific observation difficult, but your life much less so.

WHAT DO I NEED?
- your eyes
- swimming goggles or a diving mask
- a large volume of water (a swimming pool or a sun-kissed ocean surrounding a tropical island will show the effect at its best)

WHAT DO I DO? Don the goggles or mask, take a deep breath, and place your face in the water and look around.

Hold your hand at arm's length in front of your face and check its clarity. Now come up for air! Take the goggles off and do the same again.

WHAT WILL I SEE? If your goggles don't mist up, you'll be able to see objects underwater clearly, including your fingers. After you've removed the goggles you'll notice that everything you look at is blurred.

WHAT'S GOING ON? Light travels more slowly through water than it does through air. When light moves from one medium to another and undergoes a change in speed, the beam is bent, or refracted. The amount of refraction depends on the ratio of the speed of light through each medium. You can see this effect when you place a spoon in a glass of water. Where the spoon meets the water it appears to bend—this is refraction.

The human eye is delicately balanced to make sure that light entering the pupil is focused by the lens onto the retina at the back of the eye. But this system is optimized for light that has passed through air before reaching the eye. The eye has evolved to take account of the refraction that occurs when the light strikes the air/eye interface and not a water/eye interface.

When light passes straight from water to the eye, it is bent by a different amount, and so isn't correctly focused by the lens. Goggles restore the air/eye interface and normal sight is resumed.

Glasses and contact lenses take advantage of light refraction in different media by bending the light that passes through them before it reaches the eye, correcting imperfect vision in near- or farsighted people.

P.S. For those of you who prefer the answer in mathematical form, here goes.

The amount of refraction of the light depends on the difference between the refractive indices of the media on the two sides of the surface of the eye's cornea (in one case the media are air/cornea; in the other they are water/cornea). The bigger the difference, the more light is bent. Because the refractive indices of air, water, and the cornea are 1, 1.33, and 1.38 respectively, this difference is much smaller when the eye is in contact with water than when it is in contact with air.

The refraction power P of a surface is given by:

$$P = (n(1) - n(2))/R$$

Where $n(1)$ and $n(2)$ are the refractive indices of the cornea and the medium outside it respectively, and R is the radius of curvature of the cornea. P is measured in diopters. A diopter is the unit of refractive power, equal to the reciprocal of the focal length (in meters) of a given lens. Assuming a value of 0.008 meters for R, the refractive power in air is about 47 diopters and in water is about 6 diopters.

We can vary the focusing ability of our eyes to a degree. However, the increase that this can produce is much less than the loss of 41 diopters when the eye is in contact with water. In fact, the biggest change the eye can achieve is about 15 diopters.

Because the eye cannot refocus to compensate for

the lost 41 diopters when it is immersed in water, objects viewed without goggles will always seem blurred.

☉ HEAD TRAUMA

How much does a human head weigh?

I had the dubious pleasure of carrying out this experiment on live television. It has certainly been one of the most popular questions answered in *New Scientist* through the years, and here's the lowdown on how you can carry out the experiment yourself.

Because it involves immersing your whole head in water, make sure that there are always at least two people present and that children are supervised.

WHAT DO I NEED?
- a bucket of water filled to the brim
- a larger, empty vessel big enough to accommodate the bucket and its contents
- your head

If you don't have a larger vessel, you'll need a large measuring cup.

WHAT DO I DO? Place the full bucket inside the larger, empty vessel. Take a deep breath and lower your head, crown downward, into the full bucket until the water reaches the base of your chin. Hold your head still for as long as possible until any water ripples cease, then take it out and have a few gulps of air. Because a lot of water is splashed about while carrying out this experiment, unless you have a very large bathroom, it is probably best to wait for a warm, sunny day and then go outside.

WHAT WILL I SEE? Water will be displaced by your head, spill over the sides of the full bucket, and be collected in the larger vessel. Collect all the water that has spilled into this outer vessel and measure its volume. If you don't have a larger vessel to catch the water, let all the displaced water run away and then start to refill the bucket into which your head was placed with the measuring cup, taking note of exactly how much water you need to refill the bucket to the brim.

WHAT'S GOING ON? Measuring the weight of your head involves effectively isolating it from the rest of your body. But while decapitation has an obvious advantage, it has far too many drawbacks, not least that you will not be able to see the results of your experiment.

Fortunately, there are a number of solutions. The first is the least practical. Your neck vertebrae are responsible for holding your head's weight. If you hang upside down, the vertebrae in your neck move apart slightly because of the weight of your head pulling on them. So to weigh your head you simply lower yourself slowly onto a scale while hanging upside down. While you are doing this you need to continually observe the distance between the top vertebra of your neck, and your skull, using, say, an ultrasound scanner. The instant the top vertebra starts moving toward the skull, stop lowering and read the scales. Because your neck is not imparting any force onto your head, this isolates your head from your neck, thus giving you an accurate measure of your head's weight (other than a small amount via the elasticity of your vertebrae and the spring coefficient of the balance).

Of course, most people have neither the inclination to hang themselves upside down, nor have the use of an ultrasound scanner. Happily there is an easier way and it is surprisingly accurate.

It works because although bone and brain are constituent parts of our heads, the rest, like the remainder of our bodies, is mostly water. We know that one liter of water weighs one kilogram (about 2.2 pounds), so if we can measure the volume of water displaced by a head we can approximately work out how much that head weighs. If you displace four liters of water using the experiment above, your head weighs approximately four kilograms (about 8.8 pounds). If your measuring jug does not mark metric volumes, one quart is about 0.95 liters, which computes to about 2.1 pounds.

Tests on members of my family and colleagues at *New Scientist* suggest that the average weight of a human head is 4.25 kilograms, or about 9.4 pounds.

P.S. Readers should not be conned, as I was, into believing the experiment would only work in water at an excruciating 39.2°F (or to be precise, 4°C), the temperature at which one liter of water weighs exactly one kilogram. This is unnecessary because the volume of water displaced is the same whatever its temperature, even if its weight is less because it is warmer and less dense than it would be at 4°C. We suggest using warm water unless you want to have blue ears and a blue nose.

☉ COMPLETE COLLAPSE

Is it true you can crush a can just by heating up water inside it?

If you take great care, yes, you can, and you'll also see the awesome force that the weight of the atmosphere can exert on people and things that are on the surface of the earth. Just stand back while you are doing it.

WHAT DO I NEED?
* water
* a large metal can with a screw top (the square type with a metal carrying handle used to package motor oil or olive oil is fine; a tin with a press-fit lid also works)
* a stove or gas-fired barbecue (the *New Scientist* research team actually used a charcoal barbecue, but if you do, be prepared to wait)
* heat-resistant gloves (such as oven mitts)

For those who are wary of handling old gasoline cans and flames, you can try a version of this experiment using a plastic soda bottle with a screw cap and some hot water.

WHAT DO I DO? Head out to the backyard or other open space. Wash out the can, making sure no residue remains. If it has contained motor oil, you may need to use a strong detergent. Add enough water to cover the bottom an inch and very carefully stand the can upright over the flame with the top off. You'll probably want to put a child cordon around the area—this is very much stand-well-back territory. When the water starts to boil vigorously, don the gloves, remove the can from the heat, and screw the top firmly into position. Now wait a few seconds.

WHAT WILL I SEE? The can will begin to bend and twist and buckle, often making a lot of metallic-rending noises. It's quite dramatic, and you end up with a piece of crushed and twisted metal.

WHAT'S GOING ON? When you remove the can from the heat and replace the top tightly, the steam inside that has been rapidly expanded by the hot water begins to cool and contract. This means that the space inside the can is no longer at atmospheric pressure and can no longer support the sides of the can against the normal atmospheric pressure outside. The weight of the atmosphere at the earth's surface crushes the can.

P.S. The pressure of the atmosphere—caused by the weight of the air above—is said to be 1 atmosphere at sea level, or 14.7 pounds per square inch (101.325 kilopascals, in metric), depending on the units you use. This is a huge figure—the weight of a column of air one square meter in cross-section reaching from Earth's surface to the top of the atmosphere would be the equivalent of a mass of 10.2 tons at the surface—yet it is one that our bodies bear easily. Its immense force, however, can be seen in practice when applied to objects like this low-pressure can.

For a milder version of this experiment, you'll need to empty a plastic bottle, carefully pour hot water into it, screw on the top, and shake it. Unscrew the bottle taking great care: high pressure inside can cause hot water to spray as the cap is removed. Empty out the water, screw the top back on, and watch. After a while, the bottle's sides will begin to collapse inward. Much the same happens with the bottle as happens with the grilled can. The hot water heats up the air inside the bottle, expanding it and forcing some of it out of the bottle. When the air cools, the pressure in the bottle falls and, because the top is tightly fastened, no air can rush into the bottle to restore normal atmospheric pressure. The air pushing on the outside of the bottle therefore has higher pressure, and the bottle is crushed.

☉ MIXING MADNESS

Is it true that if you mix the right amount of cornstarch with the right amount of water you end up with a liquid that solidifies on impact?

You'd better believe it, because knowing this could save your life! A liquid that solidifies on impact sounds like a contradiction in terms, but that's exactly what you get if you mix cornstarch and water in the right quantities. Such liquids are known as dilatant materials and are utterly spooky because the harder you stir them the more solid they become. Stop stirring and you have a liquid again—and playing with these materials is positively mind-bending. What's more, should you ever find yourself stuck in quicksand, you'll be glad you read this book.

WHAT DO I NEED?
* twenty tablespoons of cornstarch
* one cup water
* a metal bowl
* a spoon
* a hammer

To take your experimenting even further, you'll need:
* a saucepan

WHAT DO I DO? Mix the cornstarch and water in a medium-sized bowl until it becomes too hard to move the spoon. Stop stirring and tip the bowl. Then stir again vigorously. You can also run your fingers through the mixture or even try to strike it with a hammer.

WHAT WILL I SEE? After you have mixed the liquid with the spoon and it has become stiff, stop stirring. Now tip the bowl and you'll notice the mixture becomes a liquid again and flows. If you stir it again, it immediately thickens once more. Try dipping your finger or the spoon into the mixture very slowly—if you are extremely gentle it will remain liquid, but pull out your finger or the spoon in a hurry and it will solidify. Drag your fingers quickly through the mixture and you'll be able to lift out a puttylike ball that you can work in your hands. But if you open your fingers and stop roiling it, it will quickly become a liquid again, so keep your hands over the bowl. Now try striking it with a hammer (this explains why you need a metal bowl). If you time the blow just right, the mixture will shatter. Best of all, the broken pieces will reliquefy and pool together like the shape-shifting creatures in *Star Trek: Deep Space Nine* or *Terminator 2*. You can even throw the material against an outside wall and watch it shatter. Bear in mind, though, that while this is fun, it's also very messy.

WHAT'S GOING ON? The mixture you have created is what is known as a dilatant material, or a shear-thickening fluid (STF). In these materials, viscosity increases as the force, or shear, on them is increased. So, as you have already discovered, the more pressure you apply, the more resistant to deformation they become. The more effort you put into stirring the mixture, the more it will resist. This is because the application of force to the material causes it to adopt a more ordered structure. Under normal conditions the particulates in the liquid are only loosely arranged, but the shock of any impact or pressure changes their alignment, locking them into place. When the stress dissipates, the material relaxes again.

Mining companies have to be aware of the dilatant properties of fluids. High-solid slurries such as mixtures of water and particles of mined kaolin or calcium carbonate being pumped away from mines and quarries have to be carefully monitored. If the quantities of water and particles hit critical limits, the pipelines can rapidly fill with solid dilatants, blocking the entire system.

Research into STFs has led to the development of "smart materials" that respond to changes in their surrounding environments. For example, military researchers are attempting to treat fabrics with STFs so that when a bullet strikes an STF-treated uniform, the uniform becomes rigid at the point of impact, and the bullet fails to penetrate it. Under normal conditions, however, the fabric would be as flexible as normal clothing.

However, cornstarch is made up of particles with a very high proportion of starch, which gives it another fascinating property: when heated it can be converted from an STF into a thixotropic fluid, a fluid that becomes more runny when stress is applied—in other words, almost the opposite of a dilatant material. (Thixotropy is one of the methods for getting ketchup out of a bottle discussed in the "Saucy Stuff" experiment in chapter 2.) Cornstarch's transformation from a dilatant to a thixotropic fluid is achieved by diluting the thick paste with more water and heating it in a saucepan. The starch particles in the cornstarch break open to release long molecules of starch, which leads to thixotropy.

P.S. Thick slurries of beach sand are often dilatant, leading to that old favorite of film directors: the victim trapped in quicksand. As the person struggles, the sand takes a firmer grip. But any student of dilatant materials and now every reader of this book knows how to save themselves and defy Hollywood legend. Gentle, sedate movements (like a slow breaststroke), not kicking legs and arms, provide the route to safety.

○ TEA-SOUPER

Why does soup cool more quickly than tea or coffee in a vacuum Thermos?

If you're trekking in the Rockies or cross-country skiing in Arctic climes, this is a serious consideration. You don't want to open your Thermos to find its contents aren't as warm as you'd banked on. So just why do watery beverages stay hotter longer than viscous ones?

WHAT DO I NEED?
- a pan of hot, thick soup (vegetable soup containing lots of pureed solid matter works best)
- two identical vacuum Thermos containers
- freshly made coffee or tea

WHAT DO I DO? Pour the soup into one Thermos, the tea or coffee into the other, and seal both. Take a sip from each Thermos after an hour and note which of the two liquids is hotter. Repeat this at hourly intervals.

WHAT WILL I SEE? You'll find that tea or coffee stays hotter for longer than the thick soup.

WHAT'S GOING ON? The link between heat (or thermal energy in transfer) and temperature is one of the trickiest concepts that students encounter when studying physics. Soup in a vacuum Thermos may start at the same temperature as the tea or coffee, but the soup's temperature falls faster for the following reasons.

When thermal energy (measured in joules) is transferred from the soup to the Thermos, the temperature of the soup falls according to the following equation:

Thermal energy transferred = mass of soup (in kg) × specific heat capacity of soup (joules per kg per °C) × change in temperature of soup (°C)

The important factor here is the specific heat capacity of the soup, which is the number of joules that are needed to change the temperature of one kilogram of soup by one degree Celsius. Water has a very large specific heat capacity: 4,200 joules per kilogram per degree Celsius. The specific heat capacities of tea and coffee are slightly lower, but fats (and the solid matter in the viscous soup) have much lower values, and hence their temperatures drop far more for the same thermal energy flow through the Thermos. Such foods also heat up and cool faster in all situations. A good example is whole milk, which boils much more rapidly than water or, indeed, skim milk does.

Water has this especially high specific heat capacity thanks to the unique characteristics of the hydrogen bond. Amazingly, you need to extract more energy from one kilogram of water than from one kilogram of any other common material to cool it by one degree Celsius. This means that wise outdoor types know that only water should be carried in a Thermos—all additives such as tea, coffee, and freeze-dried food are taken in dry form and added later. Although your soup will be less tasty than if it was made fresh at home, it will be hotter when you consume it—a vital consideration if you are experiencing extreme conditions.

Interestingly, despite its being an important constituent of many vegetable soups, the potato's high water content makes it an ideal thermal energy reservoir— hence the baked potato became so useful for warming hands on the prairie. And beware French onion soup. Its very watery base means that its specific heat capacity approaches that of water, but it is also protected by a floating, anti-evaporation film of oil. This can blister the tongue of the unwary diner while companions are already eating their third helping of viscous bouillabaisse. The good news in both cases is that if you're thinking of making soup before you set out for the day, potato or French onion soup should be your choice if you want to eat it hot.

One more confounding effect remains. If you are too lazy to stir your soup properly while it is heating, it may boil in only a few local spots rather than having an even heat of 212°F (100°C) spread throughout. If you put the soup into your Thermos at this stage, it will start off at a lower temperature than boiled tea or coffee and so become tepid even sooner.

P.S. We were delighted to hear from the Harrogate Ladies' College, who provided *New Scientist* with evidence to help us formulate our studies in this field. Under controlled conditions they measured the cooling curves, boiling points, and the specific heat of both tea and mushroom soup. Their graphs showed that soup does cool faster than tea, and their recorded specific heat figures of 2960 joules per kilogram per degree Celsius for soup versus 4185 for water easily account for this effect. Remember—real scientists experiment!

9 REPULSIVE VEGETABLES

Why do cabbage leaves repel water so well?

Use discarded leaves from the kitchen chopping board to check out which are waterproof and which are not.

WHAT DO I NEED?
- cabbage leaves
- leaves from other vegetables and backyard plants such as lettuce, cauliflower, spinach, and honeysuckle
- a supply of water (a watering can will do; heavy rainfall is better)

WHAT DO I DO? Sprinkle the leaves with water. If you know it's going to rain, line them up outside and watch how they repel or collect water.

WHAT WILL I SEE? The cabbage leaves and some others, like honeysuckle, will repel the water drops, forming beads that run off the surfaces. If you tip the leaves, they will become almost completely clear of water as it readily pours off the surface. Other leaves, such as lettuce, will allow the water to collect on their surfaces and, given enough coverage, will actually be damaged by the quantity and weight. Left out in overnight rain, cabbage leaves will have only a few drops on their surfaces, while lettuce leaves will be covered in a layer of water.

WHAT'S GOING ON? Most leaf types are coated in water-repellent waxes. However, cabbage leaves (and the leaves of other plants, including honeysuckle and nasturtium) have a super-hydrophobic surface, one with an extra layer of powdery wax that forms tiny pillars. When a droplet of water lands on such a surface, it touches only the tips of the wax pillars and not the leaf itself. This causes the water to form beads, which easily run away on anything other than a totally flat surface. Even if they sit on the surface, they can be removed simply by tipping the leaf.

This does not happen on leaves that are not super-hydrophobic. On these leaves, surface tension flattens out the water droplets and allows them to stay in their resting place.

You can remove the waxy surface from the cabbage by rubbing it. The leaf then acts like a lettuce leaf. You will also notice that the cabbage leaf seems greener where you have rubbed it because the wax scatters the light that hits it, giving cabbages their bluish white sheen.

The leaves of such plants are often close to the ground, where their surfaces pick up dust, dirt, and bacteria, so when the water rolls off their surfaces, they are effectively cleaning themselves. However, their surfaces reflect more sunlight than other leaf types, so this might explain why not all plants use this system, because it reduces their ability to photosynthesize.

☉ SOUND BITES

Can I measure the speed of sound?

You can indeed, although the experiment is a bit low-tech compared with the "Hot Chocolate" experiment to measure the speed of light in chapter 1. You'll also need access to a very large space.

WHAT DO I NEED?
- a helper capable of safely wielding a hammer
- the hammer
- a hammer-resistant and preferably resonant surface (a wall or a piece of metal or rock will do)
- a stopwatch (or a clock with a second hand)
- a very large backyard or flat open space, such as a beach or a park
- a pair of binoculars
- a very long measuring tape or a measured ball of string

To check out further properties of sound, you'll need:
- a metal coat hanger
- some string
- a spoon

WHAT DO I DO? It goes without saying that this job must be carried out by an adult, as a hammer is used.

Ask your helper to begin striking the hard object

once every second with the hammer—they'll need to use the stopwatch and maintain a regular beat. Begin walking away from your helper, looking back from time to time. When you are a few hundred meters apart, use the binoculars if necessary.

WHAT WILL I SEE? As the distance between you and your helper increases, the delay between them striking the hard object and the sound reaching your ears will become greater. Eventually, the delay will match the time between each beat and the sound will once more appear to coincide with the action of the hammer.

WHAT'S GOING ON? Sound has a relatively low speed through air, which means you can measure it if you have a large enough open space. Sound travels at 1,128 feet (344 meters) per second in dry air at 68°F. This may vary at different temperatures but not by much, so this experiment works pretty well just about anywhere.

When the sound of the hammer once again coincides with the one-second beat, you are ready to measure the speed of sound. Stop walking away at this point and measure the distance between you and the helper. You should be 1,128 feet away, or very close to that.

Of course, very few of us have such long backyards, and finding a flat open space of that length may prove troublesome. Then there will be the problem of outside noises interfering with you detecting the beat of the hammer. If this is the case, ask your helper to increase the frequency of the beats to once every half second (in which case you should find yourself about 564 feet, or 188 yards—still almost two football field lengths—from your helper when the beats again coincide with the sound), or once every quarter second if they can (when you'll be about 94 yards away).

P.S. The speed of sound is slow enough for noises to be noticeably delayed when heard from even short distances, as this experiment verifies. You've probably already experienced this when seeing a car door being slammed when you are a hundred yards or so away, or even when you have heard your own voice echoing back after shouting in a tunnel or beneath a bridge.

However, in liquids and solids, the speed of sound is noticeably increased. This is because sound transmission is propagated through neighboring atoms or molecules. The speed of sound depends on the interactions between these atoms and molecules—in effect, how often they collide. In air this is relatively low, but in seawater the speed of sound is 4,900 feet (1,500 meters) per second while in steel it reaches 16,750 feet (5,100 meters) per second.

In order to check out just how sound travels more efficiently through solids and liquids, try the following. Take a triangular metal coat hanger and tie a piece of string to each of the two corners that don't carry the hook. Then tie one of the strings to your left index finger and one to your right. Now lift the coat hanger so it is dangling, hook down, from the two strings. Ask a friend to tap the coat hanger with a spoon. You'll hear a tinkling sound much as you'd expect. Now, still holding the strings, put your fingers in your ears and get your friend to tap the coat hanger again. This time the sound will be more reminiscent of a peal of bells and much, much louder. That's because the sound is traveling through the solids and liquids of your body rather than through the air.

☉ BA(N)G

Why does mixing vinegar with baking soda produce an explosive mixture?

Like cornstarch, it's amazing how useful vinegar and baking soda are to the amateur scientist. Stand back and admire their versatility.

WHAT DO I NEED?
+ a tablespoon
+ baking soda
+ a tissue or small muslin bag
+ vinegar
+ a measuring jug or coffee mug
+ warm water
+ a clear plastic food bag
+ a tight-fitting plastic kitchen clip (or two if you have deft hands)

You'll need to make sure the bag doesn't leak, so take a big breath and blow in it to check that it doesn't deflate from a minute scratch or flaw.

WHAT DO I DO? Take everything outside—the results can be quite messy. Place two tablespoons of the baking soda inside the tissue (or muslin bag) and fold the tissue around the powder so you have a neat packet. Put ten tablespoons of vinegar (about half a mug) and five tablespoons of warm water (about a quarter of a mug) into the bag and seal it with the clip. Take the baking soda packet, open the bag, drop the packet in, and seal the bag as quickly as possible with the clip. To make sure it doesn't leak you can add a second clip, but be quick— the bag can inflate rapidly. Then shake the bag. Drop it and step back quickly.

WHAT WILL I SEE? The bag will expand then explode, often quite loudly and always messily.

WHAT'S GOING ON? Baking soda is essentially sodium bicarbonate, and this reacts with the acetic acid in vinegar to produce carbon dioxide, which takes up much more space than the powder and liquid from which it is formed. Therefore, the gas expands quickly to the point where it can no longer be contained by the sealed bag. Hence the explosion.

P.S. You do need to be careful when you are attempting this unless you want to find yourself covered in smelly liquid. The reaction is very rapid, so instead of shaking the bag you can just drop it and run after putting the baking soda packet inside.

○ FIZZY FLIGHT

Alka-Seltzer tablets can be used to make film canisters fly through the air. How?

If, like me, you've hung on to your 35mm SLR camera, you'll be well set to try this experiment. If you haven't, you'll have to get hold of some old 35mm film canisters. It's worth the hunt—or worth buying a roll of film—in order to see this splendid effect. The flying canisters move at considerable speed, so make sure everybody is standing well back.

WHAT DO I NEED?
- ◆ empty film canisters
- ◆ water
- ◆ Alka-Seltzer tablets
- ◆ outdoor space

WHAT DO I DO? Half-fill the film canister with water. Pop in the Alka-Seltzer tablet and quickly snap on the lid as firmly as possible. Shake the canister and stand it upside down in the middle of your backyard and step back quickly.

WHAT WILL I SEE? The canister will pop and fly upward, leaving the lid behind on the ground. Tests in the *New Scientist* yard saw the canister achieve heights of around two yards.

WHAT'S GOING ON? This miniature launch station takes advantage of the simple reaction that occurs when Alka-Seltzer is added to water. Alka-Seltzer is an antacid designed to combat the unpleasant effects of acid indigestion in the stomach. For this reason it contains sodium bicarbonate, an alkali (hence the "Alka" part of the brand name) that neutralizes the acid. It also contains citric acid, which when added to water reacts vigorously with the sodium bicarbonate to form carbon dioxide. This reaction is similar to the one seen in the "Ba(n)g" experiment on page 195.

The carbon dioxide bubbles form quickly inside the film canister, causing the pressure to rise until it is so great that the canister is blown apart between lid and base.

P.S. Children love this experiment. While you need to make sure they don't get their hands on the tablets or get struck by flying canisters, they can design colored pieces of card stock that can be folded into tubes and stuck around the canisters. Add a cone to the top of the tube and fins around the base and you can hold a competition to see whose rocket flies highest and furthest.

☉ BLOW OUT

How can I make a volcano in my backyard?

Not all volcanoes look like Mount Vesuvius—they come in different shapes and sizes and emit different products. You can try out two of the most common on a miniature scale. Even at this smaller level, however, it's best done outside.

WHAT DO I NEED?
- a fridge
- two bottles of carbonated beverage—the bigger the better (two-liter bottles work well), and colored drinks such as orange or cherry soda give a pleasing effect
- a funnel
- wallpaper paste powder
- outdoor, open space

WHAT DO I DO? Chill the bottles of soda to a temperature as low as possible without freezing the liquid. Take the bottles out of the fridge, open one without shaking it too much, and pour out enough to make room for the wallpaper paste powder you'll be adding to the bottle. Put the funnel into the open neck and add a few tablespoons of wallpaper paste powder until the bottle is full (you can experiment with different amounts of paste and soda). Quickly screw the cap back on. Now let both bottles warm to ambient temperature (a warm day is better for this experiment than midwinter). Shake the bottles vigorously, place them in the center of your open space about a yard apart, and quickly unscrew their caps.

WHAT WILL I SEE? The bottle that does not contain the wallpaper paste powder will spray its liquid out in a wet spray, the gas bubbling out through the liquid with very little of the soda leaving the bottle before it dies down again. The liquid that does come out will spread out in a puddle.

The bottle that contains the wallpaper paste powder, however, will produce a thick stream of gunk, oozing quickly from the bottle, building up around the base and spreading out much more slowly than the liquid around the other bottle.

WHAT'S GOING ON? First, a bit of background. You chill the drinks to ensure that the bubble-making carbon dioxide contained in sodas remains in solution—the colder the drink, the more carbon dioxide it can hold. Carbon dioxide also stays in solution when under pressure, so while the caps of the bottles remain in place, the carbon dioxide remains dissolved in the liquid. When the drinks warm up before you unscrew the caps to carry out the experiment, the carbon dioxide becomes less soluble. So, when the pressure is released as the caps are removed, it rapidly begins to bubble out of the drinks, forcing the liquid out of the bottles.

Once the bottles are opened you are, in effect, creating two types of volcano in garden-scale form. The bottle containing the wallpaper paste resembles a volcano emitting felsic lava. This is highly viscous and contains more than 63 percent silica. As the lava bubbles up inside the volcano it decompresses, just like the soda does as it leaves the once-pressurized bottle, and gas forms in the viscous fluid. Because the fluid is sticky in both felsic lava and the wallpaper paste soda, the bubbles cannot escape easily, so the volume of liquid swells and is forced up out of the bottle. Consequently, lava in this form tends to be more explosive than other types, as the gas struggles to escape the confines of the volcanic vent and the viscous liquid and can at times blast its way out. This explosive action and the thick lava give rise to the classic, cone-shaped volcanos we associate with Hollywood movies. Genuine examples include Mount Vesuvius in Italy and Mount St. Helens in Washington State.

The bottle without the wallpaper paste powder resembles a volcano emitting mafic (or basaltic) lava—so called because of its high magnesium and iron content. This lava has less than 52 percent silica and is far runnier. Because of this, the gas is able to bubble free after the lava decompresses at Earth's surface. This allows the liquid rock—like the pasteless soda—to run smoothly from its vent with little or no explosive effect. This leads to long flows of molten rock and volcanoes with shallower slopes because of the more fluid lava, accounting for those seen in Iceland and Hawaii. Nonetheless, while the volcano slopes are shallow, they can grow to great heights because they are rarely destroyed in catastrophic eruptions. In fact, the largest lava shield—as these volcanoes are called—is found in Hawaii. Known as Mauna Loa, it is seventy-five miles in diameter and rises more than five and a half miles from the seafloor to form part of Hawaii's Big Island. Even that is dwarfed by Olympus Mons on Mars, a shield volcano creating the tallest mountain in the solar system at 88,500 feet.

P.S. Many types of wallpaper paste contain a fungicide to prevent mold from growing on interior walls. Don't be tempted to drink the volcano once it has finished erupting!

◯ UP AND DOWN

How does a yo-yo yo-yo?

Indeed, how does a yo-yo do lots of different things? When you've mastered the technique of making the yo-yo travel up and down its string and read the how and why, have a go at a few of the special tricks we list at the end.

WHAT DO I NEED?

◆ a yo-yo—and a bit of practice

WHAT DO I DO? Play with the yo-yo in the normal way, making it travel up and down its string. If you've never done this before, it's quite tricky at first. If you are a novice, here are a few tips to get your experimentation under way thanks to Learn2Yo-Yo (http://www .tutorials.com/09/0915/0915.asp). Slip the middle finger of your hand through the loop, placing the loop in the center of your finger, between your first and second knuckle joints. Let the yo-yo dangle for a while to unravel any kinks in the string. Then wind the yo-yo up its string, and cup it in your palm with your middle finger placed in the string groove and your other fingers curved around the sides. Bend your arm to a right angle at the waist, with your forearm in front of you and your elbow at your side. Lift your forearm slightly, snap your wrist back, then drop your arm down to its original spot and release the yo-yo as if you're throwing it to the floor. Your hand should open up completely at this point. As the yo-yo travels down the string, lift your forearm and hand slightly in a smooth motion. This gives it extra speed. You'll feel a slight jerk of tension when it reaches the bottom. The yo-yo should then start its journey back up. As it does, drop your forearm and hand in the same easy motion, adding a bit more kick. The yo-yo should return to your hand just as your arm reaches waist level (the same position from which you released it). You can catch it here and start again, or keep your hand open and continue the downward arc of your arm, so your palm hits the yo-yo and pushes it down for another pass.

WHAT WILL I SEE? The yo-yo will travel down its string, reach the bottom, and travel back up again in apparent defiance of gravity.

WHAT'S GOING ON? If you examine a yo-yo, you'll see it consists of two heavy discs connected by a smooth axle. The string is not tied to the axle but simply looped around it. You can visualize this string as a U that starts at your finger, goes down and around the yo-yo's axle, and back up again to your finger. The string is twisted, forming a loop at the bottom so that the yo-yo can't jump out of the U. The string is, in fact, the most important part of a correctly set-up yo-yo and should always be the single U piece looped in half, and not tied to the axle as many often are.

As the yo-yo is released from your hand, its initial potential energy is converted into angular kinetic energy. It stores most of this energy, like a gyroscope, in its spin

spindle

braided string

or angular momentum. The angular momentum of a rotating object is the measure of the extent to which the object will continue to rotate about its axis unless acted upon by an external force. A good yo-yo has more mass on the outside of the disc than on the inside in order to keep it light while at the same time maximizing the angular momentum.

The end of the string loops loosely around the axle of the yo-yo, leaving the yo-yo free to spin inside the loop when it reaches the bottom of its descent. The friction between the string and the axle here is not great enough to allow the string to grab the axle and start winding itself up. However, if you give a small jerk to the string—this happens in normal yo-yo usage when the string jerks as the yo-yo reaches the bottom of its descent—this momentarily reduces the string's tension. The friction of the string against the axle is then enough to allow the string to start wrapping itself around the axle.

Once the string has made one turn around the axle, the loop starts to act as if it is attached—the outer turns of string stop the inner loop from slipping any more, and the yo-yo then begins to climb back up the string, converting its kinetic energy into potential energy.

The propensity of the yo-yo to remain spinning depends on the friction between the axle and the string. This can be adjusted by twisting the string to make it tighter or looser around the axle, or by waxing the inner axle to lubricate it. The smoother and looser it is, the easier it is to do some of the tricks we list below, but the harder it is to make the yo-yo climb back up the string.

P.S. Now that you know how to use a yo-yo and understand how it works, you may feel confident enough to try out some of these special tricks.

"Sleeping" a yo-yo involves dropping the yo-yo down its string in the normal fashion, but not pulling it back up, so that the yo-yo remains at the bottom of the string spinning freely. The key to sleeping the yo-yo is to have a loose loop around the axle. A tight loop, as described earlier, causes the string to grab the axle, making the yo-yo travel back up after it hits the bottom of its cycle. The loose loop allows the yo-yo to spin freely.

"Walking the dog" involves sleeping the yo-yo after giving it a particularly hard throw—to put lots of energy into its spin—and then allowing the spinning yo-yo to touch the ground. It will "walk" along the ground, spinning on its axle.

"Spanking the baby" is achieved by first sleeping the yo-yo and then slapping the back of the string-holding hand after a few seconds of sleep. This jars the string and, if the loop is not too loose, will cause momentarily increased friction and the string will catch the axle, returning the yo-yo back to your hand.

P.S. There's another property of a yo-yo that demonstrates how friction can act on bodies. Place your yo-yo on a horizontal surface and look at it from the side with the thread coming out of the bottom of the axle to the right. If you pull the string to the right, which way will the yo-yo move? The answer is left, right, or not at all—it depends on the angle *a* at which you pull the string.

When *a* is small and the thread is pointing upward, the thread tends to turn the yo-yo counterclockwise.

However, this is counteracted by friction: *b* is a point just to the right of the vertical center-line of the yo-yo, and frictional force acts on the yo-yo where the center-line touches the horizontal surface, creating a torque that opposes the anti-clockwise rotation so that the yo-yo rolls to the left rather than the right. When *a* is large and the thread is pointing nearer to the horizontal surface on which the yo-yo is resting, the yo-yo still wants to rotate counterclockwise. However, there is a strong lateral force—the thread—pulling the yo-yo to the right. Again, friction acts to the left of b and this time it doesn't just oppose the counterclockwise motion. Instead, in combination with the lateral force, it turns the yo-yo clockwise and the yo-yo rolls to the right.

Even more intriguingly, between these two extremes, the yo-yo will simply spin without moving one way or the other.

○ THE BUMBLE BALLOON

Why, when you blow up a balloon and then release it, does it spiral around as the air is expelled, rather than follow a straight path?

This experiment is best done outside. While testing it at home, I managed to damage a glass photograph frame and knock over a vase of flowers left over from the "Flower Power" experiment in chapter 3. Randomly released balloons and their wildly fluctuating flight paths carry more force than you might expect.

WHAT DO I NEED?
* party balloons
* strong lungs

To investigate balloon motion more thoroughly, you'll need:
* a cup
* drinking straws

If you wish to study the more counterintuitive properties of balloons, you'll also need:

- a piece of plastic or rubber tube
- a squeaking party noisemaker

WHAT DO I DO? Blow up the balloons and pinch the openings to make sure no air escapes. Hold them above your head, and release them.

WHAT WILL I SEE? The balloons will fly around in random, circular paths rather than in a straight line.

WHAT'S GOING ON? For a balloon to fly in a straight line, the direction of the jet of expelled air would have to be in line with the balloon's center of mass and its center of drag—the point where the forces resisting the balloon's forward motion are symmetrical.

If these two centers don't coincide, the center of drag should be behind the center of mass, otherwise stability is compromised. The reason arrows have feathers, or fletching, and darts have flights is to keep the two centers in line and ensure drag is at the rear of the moving projectile.

If the balloon's line of thrust does not pass through the center of mass (which is almost certain) but is in the same plane as the line joining the centers of mass and drag (which is unlikely), the balloon would travel in a circle in that plane, although the pull of gravity will ultimately force it down to the ground, especially as the air driving it forward expires. However, because these lines generally do not intersect, thrust from the balloon's opening comes at an angle to the plane of the circle, pushing the balloon into the helical, corkscrew motion you saw when carrying out the experiment. The thrust of the balloon and the air resistance to the balloon will not cancel each other out in such a situation and so a turning moment is exerted.

To see a similar effect, try pushing your two index fingers on the opposite sides of a cup. If the two fingers line up directly at opposite sides of the cup, it will remain stationary. However, if your fingers are not lined up, the cup will turn, in the same way the two forces—thrust and air resistance—acting on the balloon cause it to turn.

Balloons are not finely produced pieces of high-precision engineering, so it is likely that the line of thrust will never be near the center of drag, and the resulting torque will always make the balloon spiral wildly. However, you can aim the balloon to a certain extent by taping a nozzle to it at different angles (try cutting one from lengths of drinking straw). With trial and error, you may be able to get the balloon to fly in a reasonably straight line.

P.S. Balloons can teach us quite a bit about physics. Try this: Blow up two identical balloons to different sizes and pinch the ends so that the air does not escape. Connect their openings with a piece of plastic or rubber tube using sticky tape and the help of a friend (if it proves too difficult on your own). Release the pinched ends so air can flow freely through the tube.

Intuition tells you that the balloons will equalize in size. But they don't. Instead, the small balloon shrinks as it forces its air into the bigger one. A number of

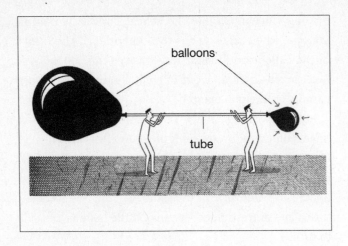

theories have been propounded to explain why this happens, and it is likely that all contain some truth. The common theme is that the air moves from the smaller balloon to the larger balloon because the pressure inside a balloon increases as its radius decreases. For a compelling demonstration, stretch the opening of an inflated balloon over a party squeaker. As the balloon deflates, the pitch emanating from the squeaker rises rather than falls, because the air is forced out ever faster. This suggests the pressure in the balloon does indeed rise as it shrinks.

Assuming we accept this, the question is: why? Some believe it's because balloons act like bubbles, which also display this relationship between pressure and radius. However, balloon rubber does not behave in the same way as a soap film. The tension in the rubber changes in a nonlinear fashion—that is, it is not a simple function of how much the rubber has been extended, and common experience supports this. The most obvious example is when you start blowing up a balloon. The initial extension of the rubber requires a lot of puff, and some people actually struggle to get past this stage. Once you do, however, the balloon inflates quite easily. This indicates that the tension in the rubber exerts a higher pressure at the initial low-extension stage than it does when it is much larger, which would account for the smaller balloon expelling its air into the larger balloon.

⊙ UP IN THE AIR

How do paper planes fly?

We've all made paper planes. Some fly quite well, while others crash. Yet even the ones that show good aeronautical ability seem to have no obvious aerodynamic features such as a lift-inducing aircraft wing. Indeed, the ones that fly best have hardly any wing area at all. So how do they fly? Are they like airplanes, or is something else at work?

WHAT DO I NEED?
- sheets of letter paper
- a windless, open space (a sports hall or similar is best, but your backyard on a still day is nearly as good)
- a decent throwing arm

WHAT DO I DO? Make the paper plane to your usual specification—most people prefer a Concorde-like dart or a delta wing with a long tail, but you probably have your own design—and launch it to see how well it flies. You can tweak the length and size of the wings or the shape of the delta wing. If you've never made a paper plane before, or even if yours is a great flier, try the design on page 214, which combines all the great characteristics of a good paper plane and, we hope, will demonstrate just how a sheet of paper can become airborne.

WHAT WILL I SEE? Your plane will fly—if it doesn't, try our design. You'll notice some loop wildly, others stall, while some fly straight but heavily toward the ground. Patience should eventually produce a version that remains pleasingly aloft for some time.

WHAT'S GOING ON? Even a stone has some very limited aerodynamic features if thrown well—consider how far a shot-putter can heave a metal ball. The mechanism by which a paper airplane flies can be simply accounted for by Newton's second law: force (in this case, lift) equals the rate of change of momentum. In simpler terms, the plane flies faster and for longer if you throw it harder, unless there is something (such as a flapping piece of paper on the nose or dangling, floppy wings) that causes drag and slows down the plane, stymieing its forward motion.

A sheet of paper cannot be thrown as far as a small stone of the same weight because of drag, which is a combination of air resistance and turbulence. Air resistance, caused by the viscosity of air, is proportional to an object's frontal cross-sectional area—or how big the plane's nose is when you look at it front-on. Turbulence is a result of twisting air currents and vortices that form around the plane. It is proportional to surface area and is reduced by a streamlined shape.

In a paper plane, long wings are useless because they are flimsy and cannot resist the bending effect involved in the lifting force. That's why short wings work better—long ones add weight and drag. Weight prevents the plane from flying far. Much of the weight of the plane comes from the wings, but the greatest lift comes from the wing area nearest the center line of the plane, so the solution is to have very small wings.

The other key factor to consider is the angle at which the plane sits as it moves forward. Air needs to strike the underside of the wings and be deflected down, ensuring a corresponding upward force on the plane. If it does not sit at an angle to the flight direction with its nose pointing slightly up, it will not generate lift. In a standard paper dart, most of the weight is at the back, which means the back drops, the nose rises, and we have lift. This lift balances the weight, and the paper dart flies. That's why—unless this lift is too effective, forcing the plane to shoot rapidly upward and stall—paper darts are better fliers than paper aircraft that actually look like real planes.

P.S. When researching what makes paper planes different from real aircraft, *New Scientist* was delighted to hear from reader Mark Wareing, whose design for a paper plane, along with his interesting explanation, follows. In tests in the *New Scientist* office using Mark's design alongside a variety of others, we found that his design stayed aloft for around 25 percent longer than more familiar paper plane configurations. We also annoyed a lot of people who were trying to work.

The ideal shape is long and thin, tapered in the front and rear like an arrow, with a small surface area for its given weight. Folding a sheet of paper into this general shape will increase the distance it can be thrown.

In addition, if the paper is folded to produce a flat, horizontal lower surface, lift will support the plane as it descends and increase the distance it can travel. As we noted above, lift is produced as air strikes the underbelly of the plane as it glides with a slight nose-up attitude.

In a conventional airplane, long, narrow wings near the middle of the body are balanced by small horizontal wings near the tail. Paper is too flimsy for building this kind of structure and flexes, particularly during launch. This means paper wings lose their aerodynamic characteristics and cause drag. And while the traditional paper dart is a streamlined, fairly rigid delta wing that often produces a stable, gliding flight, it has relatively large wings that produce excess drag and can lack rigidity.

The diagram shows how these problems can be overcome. Smaller, more robust lifting surfaces in the form of a delta wing produced from two or more layers of folded paper retain their shape better during a high-speed launch. They can be folded with a slight convex curve on the upper surface to increase lift in the way a real airplane's wings are designed. Additionally, a tightly folded structure adds mass to increase the launch momentum without significantly increasing drag.

If this delta wing is moved to the front and the lift it creates is balanced by a long, flat body folded into a V-shape toward the tail to prevent sideways movement (or yawing) during flight, the best characteristics of the paper plane can be combined into one design.

Finally, a flap of folded paper projecting down from under the nose of the plane can prove useful as a handhold for launching.

☉ CASTLES OF SAND

Why does damp sand make better sandcastles than dry sand?

The answer seems obvious: surely water just makes the sand stickier. But there's more to it than that. Try building a few sandcastles and see what happens. Obviously, you can try this out at home, but the experiment is much better if carried out on a beach, under a blue sky, with the waves lapping at your toes and a plentiful supply of ice cream.

WHAT DO I NEED?
- sand
- water
- buckets (preferably ones with molds shaped like castles)
- shovels
- ice cream (optional)

WHAT DO I DO? Make sure that some of the sand is damp (you can try out different proportions of sand/water mix) and some is dry. Pile dry sand into one bucket using your shovel and damp sand into another, then pat the sand down into the buckets to level it off. Upturn the buckets, pat them gently on their undersides, and remove them from the sandcastles.

WHAT WILL I SEE? The damp sandcastle will retain its shape while the dry sandcastle will crumble.

WHAT'S GOING ON? To understand this problem you need to understand two forces at play: friction and surface tension.

If you place a small block of wood on a table you can slide it around easily, but if you put a heavy weight on top of the block you will need to apply a much greater force to move the block—not just to get the extra mass moving, but also to overcome the increased friction. The bigger the force holding the block against the table, the bigger the force needed to make it slide.

Sand consists of lots of small grains of rock that can slide over one another. If the forces pushing the sand grains together are small, then the grains slide easily. This is the case with the dry sand in the bucket. The sand grains are only pushed together by the weight of any grains above them, so when you tip the bucket upside down they slide easily until they form a conical pile.

When the sand is damp, each grain is coated with a thin film of water, which tends to collect at the points where the grains touch. Surface tension acts at the surface of the water, producing the same result as if the water were covered by a stretched skin that is always in tension. You can see another example of surface tension if you fill a glass of water just a bit too full—a little water can be held in place a millimeter or two above the rim of the glass by surface tension (this kind of surface tension factored in the "Floaters" experiment in chapter 2).

Where the water droplets adhere to the sand grains, the tension is applied to each grain, and this effectively pushes each one against its neighbor, providing quite a strong force between them even if there is no weight of sand above. The effect of this force, related to capillary action, is enough to provide plenty of friction so that your sandcastle stays together.

As you start adding water to the sand, it forms pendular bridges between the sand grains, which hold them together, thanks to surface tension and friction.

These forces acting together can resist gravity and prevent the walls of the sandcastle from collapsing.

The surfaces of these liquid bridges are concave. This generates further capillary action, or suction, that also helps to hold the grains together. If you add a little more water to the sand, the pendular bridges start to merge, and the sand/water mix passes through the funicular state to reach the capillary state named after this force.

States of saturation of liquid-bound powders

pendular   funicular   capillary   droplet

sand castle

In this state, the concave liquid surfaces continue to generate a capillary action, which holds the sand grains together.

However, if you continue adding water you reach a point where the surface curvature of the liquid becomes convex rather than concave, and the capillary action disappears. This is known as the droplet state. The water no longer creates any attractive force between the particles, and the walls of the castle begin to slump and flow as a liquid slurry.

Interestingly, if you try to make your sandcastle entirely underwater with completely wet sand, there are no surfaces between water and air in the sand mix, so there are no surface tension forces holding the grains together, and the castle will collapse into a conical pile. This proves that it is not simply the water sticking the grains together as you might assume but actually a combination of friction, surface tension, and capillary action.

P.S. These liquid forces are used commercially to agglomerate or granulate many fine powders, such as dishwashing detergents, fertilizers, and the powders you find inside medicine capsules. In these cases, when the water that has been added to these powders evaporates, salts crystallize out of solution to form permanent solid bridges between the grains that continue to hold the agglomerate together, even when dry. You could use a similar approach when building your sandcastles, using a concentrated salt solution instead of water to see if the grains making up the castle stick together even after it has dried out. Or, if you've been following our advice and carrying out this experiment on the beach, use seawater, which has a much higher salt content than tap water, and watch what happens when your damp castles dry out.

9 NOZZLE PUZZLE

Why, when you fill a bucket of water using a hose and hold the nozzle of the hose underwater close to the bottom of the bucket, is it seemingly attracted to the bottom of the bucket?

This is a weird feeling: sometimes the nozzle feels as though it is almost being tugged toward the bottom of the bucket. It's a counterintuitive and unexpected response.

WHAT DO I NEED?
- a bucket
- a garden hose with a nozzle that can control the flow of water (rather than an open-ended piece of hose) attached to a water supply

WHAT DO I DO? Start to fill the bucket via the hose with the tap turned fully on so that the pressure of water coming out of the nozzle is reasonably powerful. Place the nozzle underwater and close to the bottom of the bucket.

WHAT WILL I SEE? The nozzle, once you hit the right position, will appear to be drawn to the bottom of the bucket, rather than away from it as the recoil from the jet stream of water might suggest would happen. You may have to adjust both the nozzle to find a suitable water-stream speed and its position in relation to the bottom of the bucket in order to feel this effect.

WHAT'S GOING ON? When the hose filling the bucket is placed so that the end of the nozzle is under the water and near the bottom of the bucket, the force pulling the nozzle toward the bottom can be explained using the Bernoulli principle. This states that the faster a fluid is moving, the lower the pressure it exerts.

When the nozzle is placed near the bottom of the bucket, the restricted area through which the water can flow between the nozzle and the bottom of the bucket causes the speed of the water to increase and produce a region beneath the nozzle that is at a lower pressure than the surrounding, slower-moving, fluid. This "sucks" the nozzle downward.

P.S. While this may seem to be a tricky concept, you can show that moving a fluid lowers pressure by holding a sheet of paper up to your mouth and blowing across its top surface. The moving air causes a drop in pressure above the paper and the paper rises rather than falls, as you might have expected. In addition to the concepts discussed in the "Up in the Air" experiment on page 211, this helps to explain how aircraft fly. The wings of an airplane are shaped to cause a faster flow of air over their upper surfaces than their undersides, and this difference in pressure causes lift.

⟨ OVERREACTION

What causes the extraordinarily explosive reaction when Mentos mints are mixed with cola or another carbonated drink?

Okay, we've saved the biggest and most spectacular for last. Many people are aware of this incredible reaction thanks to viewing the astounding results on the Internet. It's a truly awesome sight and one that we simply couldn't leave out. So, stand back and prepare to be amazed.

WHAT DO I NEED?

- a two-liter bottle of cola or other carbonated beverage (preferably of the diet variety)
- a large open space
- a tube of Mentos mints

WHAT DO I DO? Open the bottle of cola, making sure it is in an open space and well positioned so that it will not fall over. Open the pack of Mentos and make sure they are all dropped into the cola at exactly the same time. This isn't easy—one strategy is to roll a tube of paper so that it holds all the Mentos vertically and fit the paper tube into the bottle neck and release them all at once; another is to place all the Mentos in a test tube or similar slim vessel, cover the neck with paper, tip the test tube upside down over the covered neck of the bottle, and pull the paper away so the Mentos all fall into the bottle at once. Whatever method you adopt, you'll have to be able to run quickly.

WHAT WILL I SEE? A volcanic eruption of cola squirting vertically out of the bottle. Some reports have recorded a frothy blast of nearly twenty feet in height. Compared to this explosion, your volcanoes in the "Blow Out" experiment on page 198 may seem puny.

WHAT'S GOING ON? There is still some debate as to what exactly causes this reaction. Obviously cola is made from—at the most basic level—phosphoric acid, sugar or an artificial sweetener, water, and carbon dioxide held in suspension. The initial theory was that the gum arabic and gelatin in Mentos break down the surface tension in the cola that normally constrains any bubbles, allowing the carbon dioxide in suspension in the drink to expand into huge gaseous bubbles and escape very quickly. A similar situation occurs when boiling rice or pasta. You'll notice that these foods will cause a pan to boil over much more readily than if it contained the water alone. This is because similar substances in the rice or pasta disrupt the mesh of water molecules at the surface of the liquid, making it easier for bubbles to form. However, other items dropped into cola—from coins to sugar—also cause it to foam, so while gum arabic may play a role, there are almost certainly other factors.

What is not in doubt is that the huge amount of gas created causes a massive increase in pressure inside the bottle, spraying the liquid out in an incredible soda eruption.

Many scientists now think that the reaction is a physical one, rather than a chemical one. Mentos themselves are covered in tiny pits, which act as nucleation sites for bubbles to form in the way that the dusty glass in the "Over the Top" experiment in chapter 3 will fizz over with poured champagne more readily than a clean glass, because the dust particles supply multiple nucleation sites. Not only that, but Mentos sink, passing through a lot of cola very quickly and allowing pressure to build dramatically.

P.S. If you want to avoid getting covered in sticky liquid, use a diet version of your drink of choice. Strangely, they seem to work better. While it has been suggested that artificial sweeteners may have something to do with this, nobody yet knows the reason.

More important, do not—repeat, *do not*—attempt to eat Mentos washed down with cola or other carbonated drinks. This would be extremely stupid . . . and scientists, generally, try not to be stupid.

WANT TO READ MORE? Visit http://www.stevespangler science.com/experiment/00000109.

For a spectacular video of multiple cola and Mentos eruptions, check out http://www.eepybird.com/exp214 .html.

⊙ ACKNOWLEDGMENTS

Most of the entries in this book began life in the popular "Last Word" column in the weekly *New Scientist* magazine, where they had been described and explained by the magazine's readers—so a big thank-you must go to them for sharing their knowledge with us. A few entries originated with this volume's best-selling forerunners, *Does Anything Eat Wasps?* and *Why Don't Penguins' Feet Freeze?* and have been adapted for you to try out at home.

The "Last Word" column has been running in *New Scientist* for thirteen years. Readers pose questions taken from things they've noticed in the world around them, and these are answered by other readers—you can check out the full archive at http://www.newscientist.com/lastword.ns and pose new questions yourself. Send us a new question or answer and you could find yourself in the next book containing the best of the "Last Word." We welcome your input at the Web address above, via e-mail at lastword@newscientist.com, or sent to our postal address at Last Word, *New Scientist*, Lacon House, 84 Theobald's Road, London, WC1X 8NS, United Kingdom.

I would like to offer special thanks to the following individuals, all of whom contributed to this book in some way, large or small. Apologies to anybody who has been inadvertently omitted:

Anil Ananthaswamy, J. K. Aronson (Department of Clinical Pharmacology, University of Oxford, UK), John Ashton, Sophy Ashworth, Jamie Au-Yeung, Anna Baddeley, Steve

Ballinger, Byron Barrett, Hugh Bellars, David Bellis, Ralph Berman, Paul Bishop, David Blake, Emma Bland, Peter Bleackley (Astronomy Group, University of Leicester, UK), Ronald Blenkinsop, Jane Blunt, Claire Bowles, Sue Ann Bowling (University of Alaska, Fairbanks), Michael Brooks (The Argonne National Laboratory, University of Chicago), Peter Brooks, Mike Brown, Peter Bursztyn, Stephen Burt, Val Byfield, Alan Calverd, J. Neil Cape, Damian Carrington, Brad Carroll (Weber State University, Utah), Julyan Cartwright (Laboratory for Crystallographic Studies in Granada, Spain), Caspar Chater (Centre for Economic Botany, Royal Botanic Gardens, Kew, UK), Stephen Claeys, Robin Clegg (Particle Physics and Astronomy Research Council, Swindon, UK), Chris Collister, Bill Crowther (Aerospace Division, School of Engineering, University of Manchester, UK), Ben Crystall, Glyn Davies, Michael Davies (University of Tasmania, Australia), Allan Deeds, Dave Duncan, Alexandra Dutton, Martin Eastwood, David Edge, Ian Evans, Leopold Faltin, Marina Fernando (Tun Abdul Razak Rubber Research Centre, Hertfordshire, UK), Tony Finn, Richard Fisher, Tony Flury, Tom Foy, Sam Franklin, Yali Friedman, Stephen C. Fry (Institute of Cell and Molecular Biology, University of Edinburgh, UK), Jessica Fryer (vegetarian chef, Canberra, Australia), Mike Gay, Alison George, Stephen Gisselbrecht, Sally Goble, Ken Goldstein, Melanie Green, Ray Hall, Richard Hann, Martin Haswell, David Hauton, Rob Hay, David Henley, Neil Henriksen (The James Young High School, Edinburgh, UK), Jesper Henson (Firefighting Department, Safety and Environment Authority, Zurich, Switzerland), Tom Hering, Angeles Hernández Y Hernández (Andalusian Institute of Earth Sciences, Granada, Spain), Peggy Ho, Richard Horton, Guy Houlsby (Department of Engineering Science, University of Oxford, UK), Shirley Hui, Hugh Hunt, Alan Hutchison, Simon Iveson, Valerie Jamieson, Matti Järvilehto (University of Oulu, Finland), Kate Johnston (Faculty of Biological Sciences, University of Leeds, UK), Nigel Jones, Liz Kane, Roger Kearsey, Gareth

Kelly (Head of Physics, Penglais School, Aberystwyth, UK), Jennifer Kelly, Guy King, James Kingsland, Susan Krafer, Learn2Yo-Yo, Michael Le Page, Meredith Lloyd-Evans, Han Ying Loke, Ben Longstaff, Thomas Lumley, Alistair Mac-Dougall (Institute of Food Research, Norwich), Peter Macgregor, William Madil, D. P. Maitland (Department of Pure and Applied Biology, University of Leeds, UK), Sally Manders, David Mann, Paul Marks, Geoffrey Martin, Sarah Marwick, Geraldine Mathieson, David May (Physics Department, Hind Leys Community College, Shepshed, Leicestershire, UK), Howard Medhurst, D. J. Mela (Institute of Food Research, UK), Peter Milroy, Alison Motluk, Oliver and Dick Nickalls, Jon Noad (Shell International Geological Exploration Team, Netherlands), Conor Nugent, Joyce O'Hare, David Oliveira (Department of Renal Medicine, St. George's Hospital Medical School, London, UK), Padraic O'Neile, Emily Owen, Mike Perkin, Andy Phelps, Simon Pierce (University of Insubria, Varese, Italy), Oreste Piro (Mediterranean Institute of Advanced Studies in Majorca, Spain), Chris Quinn, Ian Russell (Interactive Science Ltd., Derbyshire, UK), Simon Scoltock, Alan Scott, Ben Selinger (Department of Chemistry, Australian National University, ACT, Australia), Hillary Shaw, Neil Shirtcliffe (Nottingham Trent University, UK), Colin Siddons, Andrew Smith, Frances R. Smith, Gabriel Souza, Hans Starnberg, Jim Thompson, Penny Thompson, Per Thulin, Lisa Trevithick, Tom Trull (University of Tasmania, Australia), Connie Tse, Johan Uys, Alison Venugoban, Ana Villacampa (Lawrence Livermore National Laboratory, California), Philip Ward and Class 7C (High Storrs School, Sheffield, UK), M. V. Wareing, Alan Watson (Cardiff University, UK), Peter Webber, Tony Weighill, Andy Wells, Richard Williams, Hugh Wolfson, Roger Wong, John Worthington, Paul Wright.

If you are looking for further inspiration for experiments to carry out at home, visit the excellent Web sites at http://www.thenakedscientist.com and http://www.howstuffworks.com.

A special thank-you is due to Alun Anderson; Frazer Hudson; Lucy Middleton; Jon Richfield; Ivan Semeniuk; Jeremy Webb; the production, subbing, art, Web, press, and marketing teams of *New Scientist*; Techniquest, Cardiff, UK; and everybody at Profile Books, including Paul Forty, Andrew Franklin, and Ben Usher, for their tireless efforts in the creation of this volume.

꩜ INDEX

Figures in *italics* indicate illustrations.

◑ ABOUT THE AUTHOR

MICK O'HARE is the production editor of *New Scientist* and the editor of the international bestsellers *Does Anything Eat Wasps?* and *Why Don't Penguins' Feet Freeze?* He lives in London. *New Scientist* (http:// www.NewScientist.com) is the best-selling and fastest-growing science magazine in the world.